C0-AJQ-479

Land Use Change

Science, Policy and Management

Land Use Change

Science, Policy and Management

Edited by
Richard J. Aspinall
Michael J. Hill

CRC Press is an imprint of the
Taylor & Francis Group, an **informa** business

CRC Press
Taylor & Francis Group
6000 Broken Sound Parkway NW, Suite 300
Boca Raton, FL 33487-2742

Printed in the United States of America on acid-free paper
10 9 8 7 6 5 4 3 2 1

International Standard Book Number-13: 978-1-4200-4296-2 (Hardcover)

Library of Congress Cataloging-in-Publication Data

Land use change : science, policy, and management / Richard J. Aspinall and Michael J. Hill [editors].
p. cm.
Includes bibliographical references and index.
ISBN 978-1-4200-4296-2 (hardcover : alk. paper)
1. Land use--Environmental aspects. 2. Land use--Management. 3. Land use--Case studies. I. Aspinall, Richard J. II. Hill, Michael J. (Michael James), 1937- III. Title.

HD108.3.L362 2008
333.73'13--dc22 2007027752

Visit the Taylor & Francis Web site at
http://www.taylorandfrancis.com

and the CRC Press Web site at
http://www.crcpress.com

RJA — to Chloe, for her company

MJH — to my mother, Mavis Hill,
for her lifelong friendship

Contents

Figures

Permissions

Figure I	Potschin, M. and Haines-Young, R., "Rio+10," sustainability science and landscape ecology, *Landscape and Urban Planning*, 75, 162–174, 2006. Reprinted with permission from Elsevier.
Figure 5.2	Etter, A. et al., Regional patterns of agricultural land use and deforestation in Colombia, *Agriculture, Ecosystems & Environment*, 114, 369–386, 2006. Reprinted with permission from Elsevier.
Figure 5.3	Etter, A. et al., Regional patterns of agricultural land use and deforestation in Colombia, *Agriculture, Ecosystems & Environment*, 114, 369–386, 2006. Reprinted with permission from Elsevier.
Figure 5.4	Etter, A. et al., Characterizing a tropical deforestation wave: the Caquetá colonization front in the Colombian Amazon, *Global Change Biology*, 12, 1409–1420, 2006. Reprinted with permission from Blackwell Publishing.
Figure 5.5	Etter, A. et al., Characterizing a tropical deforestation wave: the Caquetá colonization front in the Colombian Amazon, *Global Change Biology*, 12, 1409–1420, 2006. Reprinted with permission from Blackwell Publishing.
Figure 5.6	Etter, A. et al., Modeling the conversion of Colombian lowland ecosystems since 1940: drivers, patterns and rates, *Journal of Environmental Management*, 79, 74–87, 2006. Reprinted with permission from Elsevier.
Figure 5.7	Etter, A. et al., Unplanned land clearing of Colombian rainforests: spreading like disease?, *Landscape and Urban Planning*, 77, 240–254, 2006. Reprinted with permission from Elsevier.
Table 5.1	Etter, A. et al., Modeling the conversion of Colombian lowland ecosystems since 1940: drivers, patterns and rates, *Journal of Environmental Management*, 79, 74–87, 2006. Reprinted with permission from Elsevier.
Figure 9.1a	IGBP Secretariat, *Global Land Project: Science Plan and Implementation Strategy*, GLP, IGBP Report 53/IHDP Report 19, 64 pp, 2004. Reprinted with permission from the International Geosphere Biosphere Program, Stockholm.
Figure 9.1d	Steinitz, C. et al., *Alternative Futures for Changing Landscapes: The Upper San Pedro River Basin in Arizona and Sonora*, 2003. Reprinted in modified form with permission from Island Press, Washington D.C.

Preface

There is a growing international community of scholars who work on themes and issues that are central to understanding land use change as a fundamental factor in the operation of environmental and socioeconomic systems at scales from local to global. This book presents a series of chapters that address spatial theories, methodologies, and case studies that support an integrated approach to analysis of land use change. Case studies provide a series of regional test beds for theories and methodologies, and the empirical content of the case studies allows a comparative analysis of land use change issues from diverse places. Case studies thus are an important mechanism, not only for understanding the multiscale nature and consequences of land use change and the particular history of changes in case study locations, but also for eliciting general principles and factors of importance in different socioeconomic, cultural, and environmental contexts. This generalization from case studies provides input to decision making related to possible future trajectories of change.

The chapters in this book were written to present this interaction of theories, methodologies, and case studies. Additionally, all the authors are concerned with links between science and decision making, especially in relation to policy and practice. Each author attempts to enable effective communication between the academic content of his or her work with decision makers, including those concerned with policy and those concerned with land management.

All the chapters were first presented in a paper session at the 6th Open Meeting of the International Human Dimensions of Global Environmental Change Programme in Bonn, October 2005—Spatial Theory and Methodologies for Integrated Socioeconomic and Biophysical Analysis and Modeling of Land Use Change: An International Test of Theory and Method and a Comparative Synthesis of Change at Local and Regional Scales.

Acknowledgments

We thank the authors of the chapters for their prompt responses to all our requests. We also thank Professor Roy Haines-Young and Dr. Marion Potschin, University of Nottingham, for their participation in the session at the International Human Dimensions of Global Environmental Change Programme open meeting in Bonn, Germany, in October 2005 at which the papers and ideas in this book were first presented and discussed. We also thank Tai Soda and Amber Donley from Taylor & Francis/CRC Press for their patient prompting and management during the preparation of this book.

Introduction

This book addresses spatial theories and methodologies that support an integrated approach to analysis of land use change. The work focuses on spatial representation and modeling for scientific study and development of management understanding of complex, dynamic land use systems. Case studies are used to develop specific examples, not only of change in the study areas used by the different case studies, but also to illustrate the variety and commonality of data sources, methods, and issues faced when studying and developing understanding of land use change.

Land use and land cover change reflect a variety of environmental and social factors.[1,2] This necessitates that an equally varied suite of data be used for effective analysis. Remote sensing, from both satellites and air photos, provides a central resource for study of land use and land cover change. Socioeconomic surveys and censuses provide an equally important source of data on social and economic systems. Atlases and other map sources can provide data on specific environmental and socioeconomic characteristics of an area. Similarly, household and other surveys can give information on motivations, values, behaviors, and actions of decision makers and land managers, and there are many other specific data of relevance for study of land use change. These different data do vary, however, in their availability, currency, and relevance for study of land change, and this is reflected in the variety of case studies in the published literature and the specific issues and questions they address.

Similarly, methods that are appropriate for analysis of the different data sources are varied, and study of land change typically may involve methods from remote sensing, GIS (geographic information systems), process and empirical modeling, and spatial and statistical analysis, among others, as well as methods that can address qualitative and quantitative data. The need to study land use as a coupled natural and human system adds to the complexity of methods needed since interaction of systems and integration in analysis present major methodological challenges.

Thus, the purpose of this book is to illustrate issues and opportunities for study of land change by discussing relevant science and methods and by providing a series of exemplary case studies that present approaches, data, and methods to study land use change. Various themes run through the book, including the wide use of remote sensing, GIS (often implicit and supporting analysis rather than explicit and a goal of the work), and a variety of statistical and other modeling methods aimed at understanding spatial and temporal trends, patterns, and dynamics. The central importance of understanding the social, economic, policy, cultural, and institutional contexts and influences on land change is also a recurrent theme.

A further set of issues, and a major context for interpretation of results of the chapters, is associated with sustainable development and the emergence of sustainability science.[3,4,5] Sustainable development provides a context and focus for scientific research[4,6,7] as well as a guide and impetus for civic debate and decision-making.[5,8] Potschin and Haines-Young[7] present a conceptual model (the "tongue model") of the

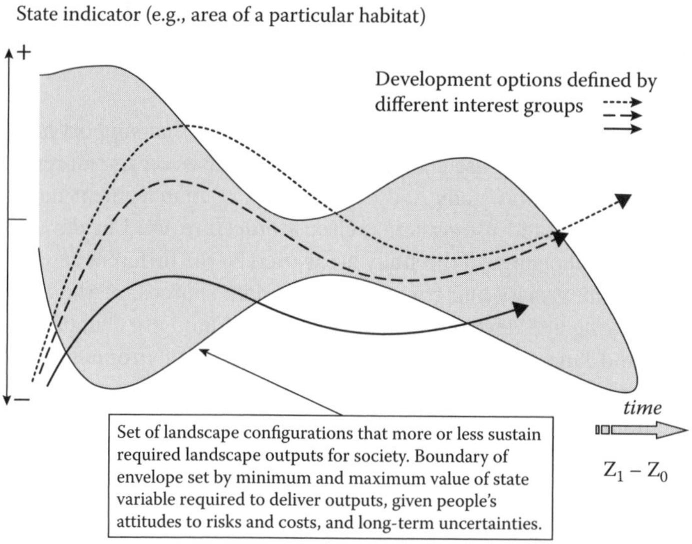

State indicator (e.g., area of a particular habitat)

+

–

Development options defined by different interest groups

time

$Z_1 - Z_0$

Set of landscape configurations that more or less sustain required landscape outputs for society. Boundary of envelope set by minimum and maximum value of state variable required to deliver outputs, given people's attitudes to risks and costs, and long-term uncertainties.

FIGURE I Conceptual model of sustainability choice space. (From Potschin and Haines-Young.[7] With permission.)

state of a landscape and its development trajectory over time (Figure I). This treats landscapes as multifunctional spaces and is founded on a paradigm of natural capital and ecosystem services.[9,10] The tongue model places boundaries on sustainability defined by a combination of biophysical limits of ecological systems in a landscape; this reflects not only environmental conditions and ecosystem services that can be achieved, but also social and cultural values, as well as costs and uncertainties that can be accepted. The tongue model[7] provides a useful framework for development and evaluation of the models and case studies of land use and land cover change presented in Chapters 3 through 8. The case studies also provide a test of the general applicability of the tongue model since they explore a diverse range of socioeconomic, cultural, and biophysical contexts for land use systems. Effective linkage of science, policy, land management practices, and decision-making related to land and ecological systems will increasingly require frameworks such as this with a focus on integration of human and natural systems that are explicitly directed at achieving consensus based on a strong scientific foundation.

The case studies also focus attention on terminology, especially related to land use and land cover. Land use refers to the social, economic, cultural, political, or other value and function of land resources. This contrasts with land cover, which refers to the biophysical properties of the land surface.[11] As such, land use and land cover, although related, are distinct from one another. The distinctions are important for understanding causes and consequences of change in land systems, as well as for mapping and other measurement of land system properties. Indeed, one of the main challenges to the study of land systems may be the tendency for conflation and confusion of the terms *land use* and *land cover.* This partly relates to wide use of remotely sensed imagery as a source of land system data but also to ambiguity and lack of distinction in definition of categories of interest. For example, the Anderson classification,[12] which is widely used as a taxonomy for land surface description based on imagery from different sources (as well as field mapping), and many other classifications confound cover and use classes within the same formal mapping legend. This may be based on apparent similarity in cover and use. For example, forestry (use) and forest (cover) may refer to the same place on the ground. It does, however, have important consequences for process-level understanding and modeling of change, presenting a limitation for recording change and consequently for understanding. Recently, efforts have been made to produce separate classifications of land use[13] and land cover,[14] and the Global Land Project[15] identifies cover and use along a continuum spanning natural and social systems.

STRUCTURE OF THE BOOK

The book is organized in nine chapters. Chapters 1 and 2 address theoretical and methodological issues and mechanisms for study of land use systems as coupled human and natural systems. Chapters 3 through 8 provide a suite of regional case studies, including a discussion of change in rapidly urbanizing areas. Collectively, these six chapters provide insight into the nature of both land use change and the diverse range of socioeconomic, cultural, and biophysical contexts of land use systems across the planet. The case study chapters provide a series of illustrations for many of the frameworks, issues, and methodologies described in the first two chapters. The empirical content of the case studies also allows a comparative analysis of land use change issues from diverse places.

Chapter 1 considers basic science questions that underpin study of land use change, as well as a suite of applied science issues that influence the utility and use of scientific information about land use change for policy and land management. The chapter focuses on issues that influence the analysis of dynamics of land change; the need for understanding the integration and feedbacks between different components of land, climate, socioeconomic, and environmental systems; resilience, vulnerability, and adaptability of land use systems; scale issues; and accuracy. The chapter also examines some issues relevant to effective communication and partnership between scientists and practitioners responsible for decision-making, either policy or management, about land use.

Full understanding of both land use change and decision-making requires an appreciation of multiple characteristics of land use systems and factors that influence change. Multicriteria decision analysis can provide a series of tools to help with both scientific study and decision-making. Chapter 2 examines the development of spatially dependent procedures and models for such multicriteria decision analysis. The approaches focus on reducing the complex responses and patterns associated with land use change to simpler meaningful metrics. These are based on use of both quantitative and qualitative measures of patterns and trends and include methods from signal processing, time series analysis, and analysis of spatial patterns.

The next six chapters present a series of case studies that are exemplars of different aspects of land use science. In Chapter 3 Byron and Lesslie explore interactions between people and the environment as a critical component in regional- and catchment-scale natural resource management. The chapter uses a case study from

Australia to bring data on the environment, from a time series of orthophotos, and social systems, from a property survey, together. This integrated analysis is complemented by use of spatially referenced survey data to understand how land managers' values, perceptions, and practices relate to the biophysical environments they manage.

Chapter 4, by Babigumira, Müller, and Angelsen, links deforestation in western Uganda in the 1990s to the socioeconomic, spatial, and institutional contexts within which it occurred. The authors develop an empirical model that integrates socioeconomic data from a national census with spatial data derived from remote sensing. The socioeconomic survey informs on poverty and economic opportunities, whereas the remotely sensed data represent the costs and feasibility of forests.

Chapter 5, by Etter and McAlpine, examines the patterns, processes, and drivers of unplanned land cover change in Colombia as representative of change in the tropics. Statistical modeling is used to predict changes in forest cover at local, regional, and national levels, over times ranging from a decade to a century. Explanatory variables include both biophysical and socioeconomic data, and these are obtained from a range of sources, including remote sensing, maps, and surveys.

Chapter 6, by Crews-Meyer, is a case study of land change in northeast Thailand that uses a time series of satellite-derived data within a longitudinal approach, panel analysis, for modeling temporal dynamics. The approach also draws on landscape ecology to emphasize the importance of the spatial scale of observation on the inference of process and attempts to examine changes in landscape composition and configuration.

Chapter 7, by Millington and Bradley, uses a case study of deforestation associated with planned colonization schemes in the Amazon Basin to develop a detailed—thick—understanding of deforestation. They argue that the impacts of roads on forest fragmentation are agents of deforestation at one scale only and that at another scale the pattern of property ownership represents that scale at which land owners make decisions about forest clearance and regrowth as household responses to economic and policy signals.

Chapter 8, by Fragkias and Seto, discusses issues of urban land use change modeling and explores the intersection of land use modeling with urban policy-making at different scales. The work also concentrates on the effects of uncertainties in data sources and reviews a predictive model of rapid urban transformation that extends a standard modeling approach to provide a policy-making framework that explicitly reduces uncertainty. Chinese cities are used as a case study as important exemplars of developing world cities.

Chapter 9 reviews findings of the full set of chapters and case studies and points to the potential developments in spatial methods needed to advance integrated modeling of land use change and improve linkage between scientific study of land use change with policy and management decision-making at regional, national, and international scales.

Richard J. Aspinall
Michal J. Hill

REFERENCES

1. Geist, H., and Lambin, E. F. Proximate causes and underlying driving forces of tropical deforestation. *Bioscience* 52(2), 143–150, 2002.
2. Geist, H. J., and Lambin, E. F. Dynamic causal patterns of desertification. *Bioscience* 54(9), 817–829, 2004.
3. Kates, R. W., *Sustainability Science. Research and Assessment Systems for Sustainability Program.* Discussion Paper 2000-33. Belfer Center for Science and International Affairs, Kennedy School of Government, Harvard University, Cambridge MA, 2000. Available at http://ksgnotes1.harvard.edu/BCSIA/sust.nsf/pubs/pub7/$File/2000-33.pdf.
4. Kates, R. W. et al. Sustainability science. *Science* 292, 641–642, 2001.
5. Clark, W. C., and Dickson, N. M. Sustainability science: The emerging research program. *Proceedings of the National Academy of Sciences* 100(14), 8059–8061, 2003.
6. Haines-Young, R. Sustainable development and sustainable landscapes: defining a new paradigm for landscape ecology. *Fennia* 178(1), 7–14, 2000.
7. Potschin, M., and Haines-Young, R. "Rio+10," sustainability science and landscape Ecology, *Landscape and Urban Planning* 75, 162–174, 2006.
8. Cash, D. W. et al. Knowledge systems for sustainable development. *Proceedings of the National Academy of Sciences* 100(14), 8086–8091, 2003.
9. Costanza, R., and Daly, H. E. Natural capital and sustainable development. *Conservation Biology* 6(1), 37–46, 1992.
10. Daily, G. C. *Nature's Services: Societal Dependence on Natural Ecosystems.* Washington, DC: Island Press, 392 pp. 1997.
11. Comber, A. J., Fisher, P. F., and Wadsworth, R. A. What is land cover? *Environment and Planning B: Planning and Design* 32, 199–209, 2005.
12. Anderson, J. R. et al. *A Land-Use and Land Cover Classification System for Use with Remote Sensor Data.* Geological Survey Professional Paper No. 964. Washington, DC: U.S. Government Printing Office, 1976.
13. Jansen, L. J. M. Harmonization of land use class sets to facilitate compatibility and comparability of data across space and time. *Journal of Land Use Science* 1(2–4), 127–156, 2006.
14. Herold, M. et al. Evolving standards in land cover characterization. *Journal of Land Use Science* 1(2–4), 157–168, 2006.
15. *GLP Science Plan and Implementation Strategy.* IGBP Report No 53, IHDP Report No 19. Stockholm, IGBP Secretariat. 64 pp. 2005.

Editors

Richard J. Aspinall received a BSc (Hons) from the University of Birmingham and a PhD from the University of Hull, both in geography. He has worked in the United Kingdom including 10 years at the Macaulay Land Use Research Institute in Scotland, and in the United States in the Department of Earth Sciences at Montana State University, as chair of the School of Geographical Sciences at Arizona State University, and as program director for geography and regional science at the U.S. National Science Foundation. He is now professor and chief executive at the Macaulay Institute in Scotland, an interdisciplinary research institute addressing sustainable development and land use. His research interests are in the areas of land use and land cover change, analysis, and modeling of coupled natural and human systems, and methods and applications in environmental geography including GIS, landscape ecology, biogeography, geomorphology, and hydrology.

Michael J. Hill received the BAgrSci and MAgrSci from LaTrobe University, Bundoora, Australia, and a PhD from the University of Sydney, Australia, in 1985. He spent 12 years in the CSIRO Division of Animal Production and then 6 years in the Bureau of Rural Sciences in the Department of Agriculture, Fisheries and Forestry of the Australian government where he carried out research in and contributed to the management of the Co-operative Research Centre for Greenhouse Accounting. In 2006, he became professor of earth systems science in the Department of Earth Systems Science and Policy at the University of North Dakota, Grand Forks. He has a background in grassland agronomy, but has been working with spatial information and remote sensing of land systems for the past 12 to 15 years. He has published widely on agronomy, ecology, biogeography, production of grasslands, and radar, multispectral, and hyperspectral remote sensing of grasslands, and more recently has been involved in the development of scenario analysis models for assessment of carbon dynamics in agricultural and rangeland systems. His current interests are in the use of MODIS land product data in model-data assimilation, application of quantitative information from hyperspectral and multiangle imaging to vegetation description, multicriteria and decision frameworks for coupled human environment systems, and methods and approaches to application of spatial data for land use management.

Contributors

Arild Angelsen

Arild Angelsen is an associate professor of economics in the Department of Economics and Resource Management at the Norwegian University of Life Sciences. His research interests are in economic analysis and assessment of projects and policies in developing countries, particularly within agriculture and forestry, and related to poverty, environmental effects, and use of natural resources.

Richard J. Aspinall

Richard J. Aspinall is professor and chief executive at the Macaulay Institute, an interdisciplinary research institute addressing sustainable development and land use. His research interests are in the areas of land use and land cover change, analysis, and modeling of coupled natural and human systems, and methods and applications in environmental geography including GIS, landscape ecology, biogeography, geomorphology, and hydrology.

Ronnie Babigumira

Ronnie Babigumira is a PhD student in the Department of Economics and Resource Management at the Norwegian University of Life Sciences. His research interest is in land use change in Africa.

Andrew V. Bradley

Andrew V. Bradley is a research scientist at the Natural Environmental Research Council (NERC) Centre for Ecology and Hydrology (CEH) at Monks Wood, United Kingdom. His research interests focus on understanding socioeconomic drivers of forest loss, forest fragmentation, and agricultural land use and land cover change.

Ian Byron

Ian Byron's research interests include analysis of sociological survey data and integration of survey results with biophysical spatial data. He co-authored the chapter while a social scientist with the Bureau of Rural Sciences, a science-policy agency within the Australian Department of Agriculture, Fisheries and Forestry.

Kelley A. Crews

Kelley A. Crews is associate professor of geography and the environment at the University of Texas, Austin, where she also directs the GIScience Center. Her thematic research interests include remote sensing, population–environment interactions, landscape ecology, and policy analysis. Geographically her work focuses on tropical and subtropical forest/savanna/wetland ecotones in the western Amazon, the Okavango Delta of Botswana, and Northeast Thailand.

Andres Etter

Andres Etter is a professor in the Department of Ecology and Territory of the School of Environmental and Rural Studies at Javeriana University (Bogotá-Colombia). He has 20 years' experience in landscape ecological mapping and research in the Colombian Amazon forests, Orinoco savannas, and Andean Montane forest regions integrating biophysical, socioeconomic, and historic data, using remote sensing and GIS. His current research deals with the modeling and understanding of land use change processes aimed at more informed and dynamic conservation planning processes.

Michail Fragkias

Michail Fragkias is the executive officer of the IHDP Urbanization and Global Environmental Change Core Project hosted by the Global Institute of Sustainability at Arizona State University. His interests focus on urban land use change and the interaction of urban spatial structure with the environment. He employs (spatial) statistical analysis, simulations, and geographical information systems (GIS) to study the significance of social, economic, and political drivers of urban land use change in China. He completed his undergraduate studies in economics at the National University of Athens in Greece and his MA and PhD in economics at Clark University in Massachusetts. He co-authored the present chapter while being a post-doctoral scholar at the Center for Environmental Science and Policy (CESP) at the Freeman Spogli Institute for International Studies (FSI) at Stanford University.

Michael J. Hill

Michael J. Hill is a professor of earth systems science in the Department of Earth Systems Science and Policy at the University of North Dakota. His research interests include hyperspectral remote sensing, biogeochemical processes, and land use change in savanna systems, and analysis of coupled human environment systems using spatial multicriteria analysis and spatial analysis methods.

Robert Lesslie

Rob Lesslie is a principal research scientist in the Bureau of Rural Sciences, a science policy bureau with the Australian government's Department of Agriculture, Fisheries and Forestry. He is an ecologist by training and retains a keen interest in landscape ecology. He currently manages the national- and catchment-scale land use mapping project for Australia, which includes work on land management practices. He has been developing multicriteria software and has a specific interest in arid lands and rangelands.

Clive McAlpine

Clive McAlpine is a senior research fellow with the Centre for Remote Sensing and Spatial Analysis with the School of Geography, Planning and Architecture, the University of Queensland, Brisbane. His research interests are in landscape ecology, biodiversity conservation, land cover change modeling, and the climate impacts of land cover change.

Andrew C. Millington

Andrew C. Millington is a professor in the Department of Geography at Texas A&M University. He has previously worked in three other geography departments: he was formerly professor and departmental chair at the University of Leicester, England; reader in geography at the University of Reading, and lecturer at Fourah Bay College (part of the University of Sierra Leone). His research focuses on the impacts of human and institutional agents on spatiotemporal patterns of land use and land cover change and the impacts of land use and land cover change on biological phenomena. He uses hybrid methodologies from GIScience, environmental science, ecology, and social science to research these phenomena. His current research includes analyzing the effects of policy initiatives on land use and land cover change in Argentina, Bolivia, and Peru, and he is initiating work on land use and land cover change in forest lands in Texas. He received his BSc in geography and geology from Hull University, his MA in geography from the University of Colorado at Boulder, and his DPhil in geography from Sussex University.

Daniel Müller

Daniel Müller is an agricultural economist now working in the Department of Economic and Technological Change, Center for Development Research at the University of Bonn. He contributed to this book during a postdoctoral appointment at Humboldt University. His research interests are in development and natural resource economics, patterns, and process in land cover and land use change, econometric analysis, and spatial data analysis.

Karen C. Seto

Karen C. Seto is an assistant professor in the Department of Geological and Environmental Sciences and a fellow at the Woods Institute for the Environment at Stanford University. Her research focuses on monitoring and forecasting land use change, especially urban growth in Asia. Her current research efforts include analyzing the effects of policy reforms on urban growth in China, India, and Vietnam. She is the Remote Sensing Thematic Leader for the World Conservation Union's (IUCN) Commission on Ecosystem Management and is a recipient of the NASA New Investigator Program in Earth Science Award and an NSF Faculty Early Career Development (CAREER) Award. She received her BA in political science from the University of California at Santa Barbara, her MA in international relations, resource and environmental management from Boston University, and her PhD in geography from Boston University.

Part I

Theory and Methodology

1 Basic and Applied Land Use Science

Richard J. Aspinall

CONTENTS

1.1 INTRODUCTION

Land use science can be defined as an inclusive, interdisciplinary subject that focuses on material related to the nature of land use and land cover, their changes over space and time, and the social, economic, cultural, political, decision-making, environmental, and ecological processes that produce these patterns and changes.[1] A variety of theories, methodologies, and technologies underpin research on land use science, and, consequently, a number of basic and applied science themes that are characteristic of land use research can be identified. These reflect the interdisciplinary and integrated analysis required to comprehend land use, as well as the role and importance of land use, land use change, and land management and policy, and the importance of land use for sustainability.[2] Land use is also considered a central part of the functioning of the Earth system[3] as well as reflecting human interactions with the environment at scales from local to global.

Basic science questions in land use science include those that focus on (a) dynamics of change in space and time; (b) integration and feedbacks between landscape, climate, socioeconomic, and ecological systems (c) resilience, vulnerability, and adaptability of coupled natural and human systems (d) scale issues and (e) accuracy. Applied science addresses policy and management questions in land use science including (a) addressing evolving public and private land management issues and decisions; (b) interpretation and communication of scientific knowledge for adaptive management of change in land use systems; and (c) human and environmental responses to change. The applied issues also should be set against a need for explicit management of uncertainties. This will include definition of the limits of applicability of change projections and other analyses, particularly as translated into decision support and participatory approaches. The need and role for spatially integrated dynamic models of coupled natural and human systems in the contexts of study and management of land use change underpin this discussion.

1.1.1 Theoretical Foundations

There has been some discussion of the potential and need for an integrated, or overarching, theory for land change. Lambin and colleagues[4] note three requirements for an overarching theory: (a) to engage the behavior of people and society and reciprocal interaction with land use, (b) to be multilevel with respect to both people and the environment, and (c) to be multitemporal in order to include both the current and past contexts in which land, people, and environment interact. Integrated study of land use and land cover changes typically is interdisciplinary or multidisciplinary in approach and thus involves theories from multiple participating disciplines.[4] The practical needs of interdisciplinary research have led empirical case studies to use a variety of mechanisms for encouraging dialogue between disciplines, including a range of integrating frameworks, most based on some form of systems representation.[5,6] Empirical studies also recognize some qualities of land systems that are common across case studies, and these suggest characteristics that a theory of land change needs to be able to incorporate:

(a) Complex causes, processes, and impacts of change[7]
(b) Differences and inter-relationships between land use and land cover[8,9,10]
(c) Interaction of socioeconomic and biophysical processes[11,12,13,14]
(d) Multiple spatial and temporal scales at which processes operate[11,15,16]
(e) Interaction across multiple organizational levels[17]
(f) Feedbacks and connections in both social and geographical spaces
(g) Multiple links between people and land[7]
(h) Influence of social, historical, and geographical context on land use change[18]
(i) Importance of individual, social, demographic, economic, political, and cultural factors in decision making[19,20]
(j) Combined use of qualitative and quantitative data and methods[21,22]

Review of multiple case studies in meta-analyses has also provided insight into land change, providing generalization about factors that lead to change. Examining 152 published case studies of tropical deforestation[19] and 132 case studies of desertification[20] Geist and Lambin identified a relatively small set of underlying causes common to the land use changes observed in different regions and places. These underlying causes are described as proximate, having apparent immediate impact on change, and ultimate, which represent fundamental causes of change.[23] There are five broad groups of underlying factors common to both sets of case studies: demographic, economic, technological, policy and institutional, and cultural; desertification also included climatic factors. Proximate causes of change common to both tropical deforestation and desertification included infrastructure extension, agricultural activities and expansion, and wood extraction; increased aridity also was a proximate cause for desertification. A meta-analysis of 91 published case studies of agricultural land intensification in the tropics,[24] intended as a companion to the meta-analyses of tropical deforestation and desertification, used the same factors as Geist and Lambin's studies and recorded a very detailed and varied list of processes associated with agricultural intensification. The main factors identified were demographic, market, and institutional, particularly property regimes. The most common processes of agricultural intensification in the tropics included adoption of new crops, planting of trees, and development of horticulture.[24]

These concerns for development and use of theory, and for improving understanding of social and natural processes, as well as their interaction, in study of land use, provide a guide for case studies and attempts at integration and synthesis across case studies. In the remainder of this chapter I discuss some basic and applied science issues that may help not only the process of studying land use and change, but also the communication and involvement of a wide variety of interested parties, including decision makers and land managers, in both the conduct of research and its implications for land management and policy.

1.2 BASIC SCIENCE

1.2.1 Dynamics of Change in Space and Time

Land use and land cover changes are inherently spatial and dynamic. The magnitude and impact of changes in land use and land cover are such that land use change is recognized as a change that is global in extent and impact.[25,26,27] Land use dynamics is also recognized as one of the grand challenges in environmental science.[28]

A variety of land use and land cover change projects have measured and monitored change in land use and land cover, particularly for periods over the past 30 years, using satellite imagery,[29,30,31] but also for periods up to the past several hundred years.[32,33] These individual projects have not only identified common land use and land cover transitions but also raised awareness and interest in the processes that produce change. For example, the International Human Dimensions Programme on Global Environmental Change/International Geosphere-Biosphere Programme Land-Use and Land-Cover Change (IHDP/IGBP LUCC) program,[6] an international program running from 1995 to 2005, coordinated a broad network of local, regional, and continental scale projects examining land cover and land use change.[14] The task of establishing cause and effect in relation to land use change dynamics faces many of the challenges of other empirical field-based sciences that rely on observation of Earth surface phenomena, such as geology.[34]

Study of land use dynamics is further complicated by a variety of time-related factors. Land use systems, as well as the underlying factors and processes, themselves may change through time. This produces a variety of path dependence[35] and legacy effects,[36] resulting in land use patterns and systems that may reflect a variety of not only contemporary processes, but also processes and responses to historic drivers of change. Additionally, land cover change involves both conversion and modification of cover[37] and may be gradual or episodic.[7] Typically change through time, especially for spatial models, is studied quantitatively for a place with a series of snapshots of land cover (sometimes treated as equivalent to or interchangeable with land use). This may not only underestimate the extent of change but also fail to capture whether changes are gradual or episodic. Measurement, analysis, and modeling of land use systems need frameworks, tools, and methods that help to separate multiple influences and asynchronous causes of change in land use system dynamics, as well as provide an improved ability to detect a greater range of types and rates of change. Crews' research[38] (Chapter 6) uses panel analysis to focus on the longitudinal (time) sequence of change.

The use of snapshots of land cover to study change and develop quantitative models may also ignore the rich source of insight and methods from a study of historical development in that place.[7,36] For example, environmental history, narrative, and storytelling all offer insight into change through time.[39] Combining quantitative approaches with environmental history would offer advantages for study of change in spatial pattern and temporal dynamics of land systems, and for understanding the roles of multiple causes and processes of change.

A range of other issues needs to be addressed to support research on dynamics of land change, especially if the goals are improved understanding of processes and changes produced and better modeling of place-based change in land systems. Specifically, there is a need for new experimental and observational designs as well as research protocols that support quantitative and qualitative analysis of change. This will enable closer and direct comparison of results from different case studies, leading to improved meta-analyses and synthesis between case studies. There is also a need for research into how spatial and temporal dynamics might best be represented to support quantitative and qualitative analysis. Although geographical information systems (GIS) and remote sensing provide mechanisms for recording, representation, measurement, and analysis of the spatial structure of properties of the land surface, GIS is still poorly developed for representation and analysis of spatiotemporal data. Improved GIS data structures for land cover and use data that explicitly incorporate time are needed. This will be a benefit not only for new forms of analysis, but it also will provide an empirical foundation for the broad range of novel tools, including agent-based modeling[40] and cellular automata[41,42] that have added to our ability to represent and model social and spatial processes that are central to a more full understanding of land system dynamics. Regular, repeated remotely sensed measurements are also needed to help develop the spatial and temporal history of land use and land cover as a precursor to improved dynamic spatial models of change.[43,44]

1.2.2 Integration and Feedbacks between Landscape, Climate, Socioeconomic, and Ecological Systems

Land systems are increasingly being treated as exemplars of coupled natural and human systems (frequently socioecological systems,[45] which are considered here to be exactly synonymous). This systems-based view of land recognizes that land use and land cover systems are defined at the interface of natural/biophysical and human systems. Indeed, the IHDP/IGBP Global Land Project establishes a framework for combined study of land change, ecosystem services, and sustainability within an Earth system science context through exploring land use and land cover as representing a continuum between socioeconomic (use) and biophysical (cover) systems.[3] This attempt at a coupled and holistic view of land systems presents a rich framework for understanding multidirectional impacts and feedbacks between elements of land systems and the many underlying factors and processes that operate to shape and change land. For example, not only do land cover and land use reflect environmental systems and the opportunities and constraints they provide, but land cover and land use also have a strong and direct influence on climate and other environmental change and must be included in new models of the Earth's climate system.[27,46] Similarly, changes in land systems reflect socioeconomic processes that operate at a very wide range of spatial and temporal scales, including globalization, trade and markets, policy and land management decisions at the national, regional, local, or household/individual level.

There have been several frameworks for study of land systems as exemplars of coupled natural and human systems. For example, Machlis and colleagues[5] describe a human ecosystem model that attempts to link social systems and natural systems through a focus on social systems linked to natural, cultural, and socioeconomics as critical resources. This explicitly includes social structures, but does not integrate an ecological systems view, and does not address process-level understanding. Despite these attempts and the claims of certain disciplines to provide an overarching integrative content, there remains a need for better conceptual models of integration and feedbacks between landscapes, climate, ecological (environmental), and socioeconomic systems. Steinitz and colleagues[47] provide a research framework for landscape design based on different types of model and the questions the models answer. Again, this has many elements that suggest integration but with a focus on models and core questions that may not be fully inclusive.

In general, qualities of successful frameworks might include (a) a focus on interdisciplinary or transdisciplinary[48] approaches, (b) a theoretical grounding in a range of sciences to allow for participation from multiple disciplines, (c) being systems based, to address processes and structures, (d) being spatially and temporally explicit to allow geographic and historical contexts and contingencies to be included in understanding and analysis of dynamics, and (e) being explicit in addressing scale, including spatial, temporal, and organizational scales,[49] including multiscale effects.[50]

1.2.3 Resilience, Vulnerability, and Adaptability of Land Systems as Coupled Natural and Human Systems

In conjunction with a growing understanding of land systems as exemplars of coupled natural and human systems, there is a move to interpret land systems within the context of sustainability science.[3,51] As an integral part of this, research addresses the vulnerability (from the hazards literature), resilience (from the ecological literature), and adaptability of land use and land systems. Sustainability, vulnerability, resilience, and adaptability all have strong temporal elements that refer not only to the ability of land systems to respond to change but also to the ability of land systems to influence the responses of socioeconomic and environmental systems to change. What would be the characteristics of a resilient land system or a vulnerable land system? Can we assess whether a land system or land use is resilient or vulnerable to changes in the external driving processes (such as in international prices for crops)? Does a land system confer resilience or vulnerability to social systems? For example, are there patterns of land use and land cover, and a suite of associated land management mechanisms, that make a community resilient/vulnerable to drought, flood, or other environmental or socioeconomic change? What are the time scales for resilience/vulnerability that are most appropriate for sustainability? Land systems clearly offer the potential to explore economic, social, institutional, environmental, and ecosystem resilience, vulnerability, and sustainability with benefits to society, environment, and land.

1.2.4 Scale Issues

Scale issues are always central to discussions of land system change,[52] but their resolution is linked to the many different meanings of "scale" in the interdisciplinary and multidisciplinary communities working on land use science. Scale refers to all of (a) spatial extent and resolution, (b) time, including the duration of a study and the resolution of data snapshots, (c) taxonomic level, for example, the level of interest in land use or land cover as in the detail in a land use/cover key or in the institutional/geopolitical hierarchy of global-international-national-regional-local, or (d) other analytical dimensions used for study. Given the interdisciplinary nature of much of the research and literature on land use, some consistency and explicit reporting of the use of terminology related to scale would be valuable.[52] Gibson and colleagues[49] provide a detailed discussion of scale issues. Scale issues in land use change may also benefit from the space-time approaches developed in landscape ecology.[53,54] A systematic analysis and review of scales at which processes and responses operate in land use systems may be valuable in guiding analysis and modeling of change.

1.2.5 Uncertainty

Uncertainty underlies many aspects of land use science and is associated not only with measurement, analysis, and modeling, but also with decision making and the consequences of models for projecting current and future conditions and states of land use systems. The accuracy of models that predict change is a constant concern of land use scientists,[55] but there are also concerns related to the accuracy and uncertainties of systems models as a whole compared with the accuracy and uncertainties of component (sub-) models of land use systems. Error propagation through coupled models of land use systems as coupled human and natural systems is of particular concern and interest.

Although statistical methods for management of uncertainty associated with measurement, models, and predictions are relatively well developed, other types of uncertainty require attention. Conroy[56] describes four types of uncertainty:

1. Statistical uncertainty: reflecting inability to measure
2. Structural uncertainty: reflecting inability to describe and model system dynamics
3. Partial controllability: reflecting inability to control decisions
4. Inherent uncertainty: present in all systems (stochastic processes)

All of these aspects of uncertainty need research for effective use of models to predict change and to use models for decision support related to policy and land management. Better integration and understanding of scientific uncertainty related to decision making under conditions of uncertainty should help to enhance communication between scientists, policymakers, and land managers.

Policy-relevant models raise some special concerns beyond those of uncertainty. Lee[57] and King and Kraemer[58] have discussed requirements for models that are to be used in policy contexts. Lee[57] identifies five qualities of models themselves: (1) transparency, (2) robustness, (3) reasonable data needs, (4) appropriate spatiotemporal resolution, and (5) inclusion of enough key policy variables to allow for likely and significant policy questions to be explored, while King and Kraemer[58] concentrate on the role served by models: (1) to clarify the issues in the debate, (2) to enforce a discipline of analysis and discourse among stakeholders, and (3) to provide an interesting form of advice, mainly in the form of what not to do. Agarwal and colleagues[59] use these criteria, criteria from Veldkamp and Fresco[60] describing space, time, biophysical factors, and human factors, as well as scale and complexity of models as cross-cutting criteria to review and compare land use change models. They take a pragmatic view that land use models, although not ideal, will be good enough to be taken seriously in the policy process. They conclude by identifying a need for greater collaboration between land use modelers and policymakers. Such collaboration will need to identify the key variables and sectors of interest, scales of analysis, and change scenarios anticipated. This will also require translation between the demographic, economic, technology, institutional and policy, and cultural factors[19,20,24] used for analysis, understanding, and modeling of land use change and the specific needs of policymakers.[59] Collaboration in this form requires closer focus on applied science from land use scientists and consideration of mechanisms used for collaboration and participation between scientists with policymakers and land managers.

1.3 APPLIED SCIENCE

Land use science is also an applied science with clear links to policy and practice through decision making and other human intervention and action on land use and land cover. Three issues seem particularly important in relation to emerging trends and needs for land use science as an applied science:

1. Addressing evolving public and private land management issues and decisions
2. Interpretation and communication of scientific knowledge for adaptive management of change in land use systems
3. Understanding human and environmental responses to change

1.3.1 Addressing Evolving Public and Private Land Management Issues and Decisions

The multiscale nature and consequences of land use change present a compelling case for translating science into policy and practice. Since land use is at the interface of human and natural systems, improved understanding of the social and biophysical processes that produce change in land use systems can play a useful role in both policy and practice for both private and public land management issues and decisions. Improved understanding of land use change can provide input to evidence-based policy and be used to develop alternative scenarios for land use response to different policies.[61,62,63] This may help to inform wide participation of interested groups and individuals in debate and discussion and also help to lead to improved decision making. This may be through improved and informed consensus by helping to explore impacts and consequences of particular policies (and alternatives) or other actions. Land use change also reflects links that occur across organizational scales and thus is important in linking broader national, international, and global trends and conditions with consequent understanding and concern for livelihoods and sustainability of communities at more human local scales.

In this context, case studies of land use change can be valuable in two ways. First, they can provide generic understanding that can be used as evidence in support of discussion to produce evidence-based policy. Second, they can provide specific scenarios and impacts for an area of interest to help focus decision making in a deliberate and locally relevant manner. Thus studies of land use change can be used not only to elucidate general issues and principles concerning land use change, but also to provide insight into applied issues associated with land use and potential meaning of change for communities and places. The latter would be of value in establishing improved understanding of social feedbacks and in developing participatory processes and consultation in decision making. These opportunities and uses suggest that there are a number of important concerns that should be made explicit in case studies of land use change. For example, are case studies concerned with management and/or policy? Are stakeholders involved in the research and in what ways? What are the reactions to results and process of the research, especially from a variety of different groups? What lessons are learned from a case study about the policies and management practices that influence land use change in the study area and systems? Better understanding of these aspects of land use change case studies will help to improve links between science and practice[37] and also place land use change evidence in both a scientific and decision and management context, getting land use/systems science translated from science to management and policy, and vice versa.

1.3.2 Interpretation and Communication of Scientific Knowledge for Adaptive Management of Change in Land Use Systems

Interpretation and communication of scientific knowledge are of increasing importance across all of science. Land use–related research, notably for agriculture and forestry, has a long history of direct and strong application to land management; examples may be found in agricultural extension services or forestry worldwide. There are also examples of land use information directly guiding policy (e.g., Land Utilisation Survey in Great Britain[64]). Communication of case study content and information on land use change is no less important today, not only for input to policy and land management decisions, but also for understanding local human processes that produce many aspects of contemporary change in land use and land cover (as reflected in the generic underlying causes previously discussed). This raises the question of the extent to which case studies involve land managers and policy or other decision makers as part of the research team (transdisciplinary in the sense of Tress and colleagues[48]). Such a two-way collaboration would be beneficial both to scientists and practitioners.[65,66]

1.3.3 Human and Environmental Responses to Change

Application of land use science to understanding practical consequences of change for human and environmental systems and their component subsystems potentially provides links to a wide variety of areas of concern in environmental management and provision of ecosystem services.[67] Improved understanding and predictive and scenario-based tools also have application in land use planning and environmental management and care.[68] Key applied issues include the manner in which different institutions, society, and individuals respond to change; how environmental systems respond; space and time scales of response and consequences; development of strategies to adapt to and manage change and its impacts; and long- and short-term consequences of change and decisions.

These applied issues should be set against a need for explicit management of uncertainties, which is a recurrent theme in scientific communication. Increased awareness of uncertainty should include definition of the limits of applicability of projections of land use change and other analyses, including scenarios, particularly as translated into decision support and participatory approaches. Boundary organizations[65] and transdisciplinary[48] approaches can help this explicit management of uncertainty since a wider community is engaged in modeling and analysis, which both helps the scientists and practitioners through improved communication and greater understanding of both decision-making needs and scientific processes.

REFERENCES

1. Aspinall, R. J. Editorial. *Journal of Land Use Science* 1(1), 1–4, 2006.
2. Raquez, P., and Lambin, E. F. Conditions for a sustainable land use: Case study evidence. *Journal of Land Use Science* 1(2–4), 109–125, 2006.
3. GLP. *Science Plan and Implementation Strategy*. IGBP report No. 53, IHDP Report No. 19. Stockholm, IGBP Secretariat. 64 pp, 2005.

4. Lambin, E. F., Geist, H. J., and Rindfuss, R. R. Introduction: Local processes with global impacts. In: Lambin, E. F. and Geist, H. J., Eds. *Land-Use and Land-Cover Change: Local Processes and Global Impacts*. IGBP Series. Springer-Verlag, Berlin, 1–8, 2006.
5. Machlis, G. E., Force, J. E., and Burch, W. R. The human ecosystem, Part I: The human ecosystem as an organizing concept in ecosystem management. *Society and Natural Resources* 10, 347–367, 1997.
6. Turner, B. L. et al., eds. *Land Use and Land Cover Change: Science/Research Plan*. IGBP Report No. 35, HDP Report No. 7. Stockholm, Sweden, International Geosphere-Biosphere Programme, 132 pp, 1995.
7. Lambin, E. F., Geist, H. J., and Lepers, E. Dynamics of land-use and land-cover change in tropical regions. *Annual Review of Environment and Resources* 28, 205–241, 2003.
8. Brown, D. G., Pijanowski, B. C., and Duh, J. D. Modeling the relationships between land use and land cover on private lands in the upper midwest, USA. *Journal of Environmental Management* 59(4), 247–263, 2000.
9. Brown, D. G., and Duh, J. D. Spatial simulation for translating from land use to land cover. *International Journal of Geographical Information Science* 18(1), 35–60, 2004.
10. Xu, J. C. et al. Land-use and land-cover change and farmer vulnerability in Xishuang-banna prefecture in southwestern China. *Environmental Management* 36(3), 404–413, 2005.
11. Walsh, S. J. et al. Scale-dependent relationships between population and environment in northeastern Thailand. *Photogrammetric Engineering and Remote Sensing* 65(1), 97–105, 1999.
12. Veldkamp, A., and Verburg, P. H. Modelling land use change and environmental impact. *Journal of Environmental Management* 72(1–2), 1–3, 2004.
13. Urama, K. C. Land-use intensification and environmental degradation: Empirical evidence from irrigated and rain-fed farms in south eastern Nigeria. *Journal of Environmental Management* 75(3), 199–217, 2005.
14. Lambin, E. F., and Geist, H. J., eds. *Land-Use and Land-Cover Change: Local Processes and Global Impacts*. IGBP Series. Springer-Verlag, Berlin, 222 pp, 2006.
15. Riebsame, W. E., Gosnell, H., and Theobald, D. M. Land use and landscape change in the Colorado Mountains. 1. Theory, scale, and pattern. *Mountain Research and Development* 16(4), 395–405, 1996.
16. Veldkamp, A., and Fresco, L. O. CLUE-CR: An integrated multi-scale model to simulate land use change scenarios in Costa Rica. *Ecological Modelling* 91(1–3), 231–248, 1996.
17. Bousquet, F., and Le Page, C. Multi-agent simulations and ecosystem management: A review. *Ecological Modelling* 176(3–4), 313–332, 2004.
18. Adger, W. N. Evolution of economy and environment: An application to land use in lowland Vietnam. *Ecological Economics* 31(3), 365–379, 1999.
19. Geist, H., and Lambin, E. F. Proximate causes and underlying driving forces of tropical deforestation. *Bioscience* 52(2), 143–150, 2002.
20. Geist, H. J., and Lambin, E. F. Dynamic causal patterns of desertification. *Bioscience* 54(9), 817–829, 2004.
21. Hill, M. J. et al. Multi-criteria decision analysis in spatial decision support: The ASSESS analytic hierarchy process and the role of quantitative methods and spatially explicit analysis. *Environmental Modelling and Software* 20(7), 955–976, 2005.
22. Madsen, L. M., and Adriansen, H. K. Understanding the use of rural space: The need for multi-methods. *Journal of Rural Studies* 20(4), 485–497, 2004.
23. Geist, H. J., and Lambin, E. F. *What Drives Tropical Deforestation? A Meta-Analysis of Proximate and Underlying Causes of Deforestation Based on Subnational Case Study Evidence*. LUCC Report Series No. 4., Land-Use and Land-Cover Change

(LUCC) Project IV. International Human Dimensions Programme on Global Environmental Change (IHDP) V. International Geosphere-Biosphere Programme (IGBP), 116 pp, 2001.
24. Keys, E., and McConnell, W. J. Global change and the intensification of agriculture in the tropics. *Global Environmental Change* 15, 320–337, 2005.
25. Vitousek, P. M. Beyond global warming: Ecology and global change. *Ecology* 75, 1861–1876, 1994.
26. Vitousek, P. M. et al. Human domination of Earth's ecosystems. *Science* 277(15 July), 494–499, 1997.
27. Foley, J. A. et al. Global consequences of land use. *Science* 309, 570–574, 2005.
28. National Research Council (NRC). *Grand Challenges in Environmental Sciences.* Report from the Committee on Grand Challenges in Environmental Sciences. National Academy Press, Washington, DC. 96 pp, 2001.
29. Skole, D., and Tucker, C. Tropical deforestation and habitat fragmentation in the Amazon: Satellite data from 1978 to 1988. *Science* 260(5116), 1905–1910, 1993.
30. Brown, D. G. Land use and forest cover on private parcels in the upper midwest USA, 1970 to 1990. *Landscape Ecology* 18(8), 777–790, 2003.
31. Ferraz, S. F. D. et al. Landscape dynamics of Amazonian deforestation between 1984 and 2002 in central Rondonia, Brazil: Assessment and future scenarios. *Forest Ecology and Management* 204(1), 67–83, 2005.
32. Etter, A., and van Wyngaarden, W. Patterns of landscape transformation in Colombia, with emphasis in the Andean region. *Ambio* 29(7), 432–439, 2000.
33. Goldewijk, K. K. Estimating global land use change over the past 300 years: The HYDE Database. *Global Biogeochemical Cycles* 15(2), 417–433, 2001.
34. Schumm, S. D. *To Interpret the Earth: 10 Ways to Be Wrong.* 2nd ed. Cambridge University Press, Cambridge. 143 pp, 2006.
35. Brown, D. G. et al. Path dependence and the validation of agent-based spatial models of land use. *International Journal of Geographical Information Science* 19(2), 153–174, 2005.
36. Aspinall, R. J. Modelling land use change with generalized linear models—a multimodel analysis of change between 1860 and 2000 in Gallatin Valley, Montana. *Journal of Environmental Management* 72(1–2), 91–103, 2004.
37. Lesslie, R., Barson, M., and Smith, J. Land use information for integrated natural resources management—a coordinated national mapping program for Australia. *Journal of Land Use Science* 1(1), 45–62, 2006.
38. Crews-Meyer, K. A. Characterizing landscape dynamism using paneled-pattern metrics. *Photogrammetric Engineering and Remote Sensing* 68(10), 1031–1040, 2002.
39. Wyckoff, W., and Hansen, K. Settlement, livestock grazing and environmental change in southwest Montana, 1860–1990. *Environmental History Review* 15, 45–71, 1991.
40. Parker, D. C. et al. Multi-agent systems for the simulation of land-use and land-cover change: A review. *Annals of the Association of American Geographers* 93(2), 314–337, 2003.
41. Batty, M. Agents, cells, and cities: new representational models for simulating multiscale urban dynamics. *Environment and Planning A* 37(8), 1373–1394, 2005.
42. Lau, K. H., and Kam, B. H. A cellular automata model for urban land-use simulation. *Environment and Planning B-Planning and Design* 32(2), 247–263, 2005.
43. Turner, D. P., Ollinger, S. V., and Kimball, J. S. Integrating remote sensing and ecosystem process models for landscape- to regional-scale analysis of the carbon cycle. *Bioscience* 54(6), 573–584, 2004.
44. Tralli, D. M. et al. Satellite remote sensing of earthquake, volcano, flood, landslide and coastal inundation hazards. *ISPRS Journal of Photogrammetry and Remote Sensing* 59(4), 185–198, 2005.

45. Carpenter, S. et al. From metaphor to measurement: Resilience of what to what? *Ecosystems* 4(8), 765–781, 2001.
46. Foley, J. A. et al. Incorporating dynamic vegetation cover within global climate models. *Ecological Applications* 10(6), 1620–1632, 2000.
47. Steinitz, C. et al. *Alternative Futures for Changing Landscapes: The Upper San Pedro River Basin in Arizona and Sonora*. Washington, D.C.: Island Press. 202 pp, 2003.
48. Tress, B. et al., eds. *Interdisciplinary and Transdisciplinary Landscape Studies: Potential and Limitations,* DELTA Series 2. Wageningen, 192 pp, 2003.
49. Gibson C. C., Ostrom E., and Anh T. K. The concept of scale and the human dimensions of global change: A survey. *Ecological Economics* 32, 217–239, 2000.
50. Redman, C. L., Grove, J. M., and Kuby, L. H. Integrating social science into the long-term ecological research (LTER) network: Social dimensions of ecological change and ecological dimensions of social change. *Ecosystems* 7(2), 161–171, 2004.
51. Kates, R. W., and Parris, T. M. Long-term trends and a sustainability transition. *Proceedings of the National Academy of Sciences* 100, 8062–8067, 2003.
52. Verburg, P. H. et al. Land use change modelling: Current practice and research priorities. *GeoJournal* 61, 309–324, 2004.
53. Delcourt, H. R., Delcourt, P. A., and Webb T. A., III. Dynamic plant ecology: The spectrum of vegetation change in space and time. *Quaternary Science Reviews* 1, 153–175, 1983.
54. Delcourt, H. R., and Delcourt, P. A. Quaternary landscape ecology: Relevant scales in space and time. *Landscape Ecology* 2, 23–44, 1988.
55. Pontius, R. G., and Spencer, J. Uncertainty in extrapolations of predictive land-change models. *Environment and Planning B-Planning and Design* 32(2), 211–230, 2005.
56. Conroy, M. J. Conservation and land use decisions under uncertainty: Models, data, and adaptation. In: Hill, M. J., and Aspinall, R. J., Eds., *Spatial Information for Land Use Management*. OPA Overseas Publishing Associates, Ltd., Reading, U.K., 145–158, 2000.
57. Lee, D. B., Jr. Requiem for large-scale models. *AIP Journal* (May), 163–177, 1973.
58. King, J. L., and Kraemer, K. L. Models, facts, and the policy process: The political ecology of estimated truth. In: Goodchild, M. F., Parks, B. O., and Steyaert, L. T., eds. *Environmental Modeling with GIS*. Oxford University Press, New York, 353–360, 1993.
59. Agarwal, C. et al. *A Review and Assessment of Land-Use Change Models: Dynamics of Space, Time and Human Choice*. CIPEC Collaborative Report Series No. 1. CIPEC, Indiana University. 90 pp, 2002.
60. Veldkamp, A., and Fresco, L. O. CLUE: A conceptual model to study the conversion of land use and its effects. *Ecological Modelling* 85, 253–270, 1996.
61. Steinitz, C. et al. A delicate balance: Conservation and development scenarios for Panama's Coiba National Park. *Environment* 47(5), 24–39, 2005.
62. Berger, P. A., and Bolte, J. P. Evaluating the impact of policy options on agricultural landscapes: An alternative-futures approach. *Ecological Applications* 14(2), 342–354, 2004.
63. van Dijk, T. Scenarios of Central European land fragmentation. *Land Use Policy* 20(2), 149–158, 2003.
64. Sheail, J. Scott revisited: Post-war agriculture, planning and the British countryside. *Journal of Rural Studies* 13(4), 387–398, 1997.
65. Guston, D. H. Boundary organisations in environmental policy and science: An introduction. *Science, Technology and Human Values* 26(4), 399–408, 2001.
66. Lemos, M. C., and Morehouse, B. J. The co-production of science and policy in integrated climate assessments. *Global Environmental Change* 15, 57–68, 2005.

67. Rounsevell, M. D. A. et al. Future scenarios of European agricultural land use II. Projecting changes in cropland and grassland. *Agriculture Ecosystems and Environment* 107(2–3), 117–135, 2005.
68. Haberl, H., Wackernagel, M., and Wrbka, T. Land use and sustainability indicators. An introduction. *Land Use Policy* 21(3), 193–198, 2004.

2 Developing Spatially Dependent Procedures and Models for Multicriteria Decision Analysis

Place, Time, and Decision Making Related to Land Use Change

Michael J. Hill

CONTENTS

2.1 INTRODUCTION

Land use change occurs within a space-time domain. Frameworks for assessing appropriate land use and priorities for change must capture the complexity, reduce dimensionality, summarize a hierarchy of main effects, transfer signals and patterns, and transform information into the language of the political and economic domains,[1] yet retain the key dynamics, interactions, and subtleties. Spatial interaction, temporal cycles, responses and trends, and changes in spatial patterns through time are important sources of information for condition, planning, and predictive assessments. Spatially applied multicriteria analysis[2] enables diverse biophysical, economic, and social variables to be mapped into a standardized ranking array; used as individual indicators; combined to develop composite indexes based on objective and subjective reasoning; and used to contrast and compare hazards, risks, suitability, and new landscape compositions.[3,4,5,6] The multicriteria framework allows the combination of multi- and interdisciplinarity.[7] The system definition depends upon the purpose of the construct, scale of analysis, and set of dimensions, objectives, and criteria.[7]

When mapping both quantitative and ordinal data into factor layers, retention of, and access to, rationale and reasoning for inclusion and weighting or contribution to composites is important for maintenance of the link between the outcome of the analysis and the real or approximate data used as input. This particularly applies to spatial and temporal information. Here it is important to know what the meaning of a spatial or temporal metric might be when it is included among other data in development of an assessment to aid decision makers. The meaning has two components: (1) the first relates to the direct description of the metric such as the average patch size of remnant vegetation within a particular analytical unit, or the amplitude of the seasonal oscillation in greenness from a normalized difference vegetation index (NDVI) profile; (2) the second relates to what the metric measures in terms of influence on the target issue; for example, patch sizes greater than x indicate a higher water extraction to water recharge ratio, resulting in a lowering of the water table, or an amplitude equal to y indicates a 75% probability that the area is used for cereal cropping and hence has no water extraction capacity in summer. In the context of multicriteria analysis (MCA), assignment of meaning to spatial and temporal metrics depends on project-based research, wherein a relationship is established between some aspect of land use change or condition, or some derived property of an input variable layer, and a metric that is robust and translatable from study to study. Intrinsically, some metrics have more easily ascribed meaning than others—the meaningfulness being inversely proportion to the degree of abstraction and extent of removal from biophysical, economic, or social measures that are a directly related to the manifestation of land use change.

There is a very wide array of potential analytical adjuncts to MCA.[8] These can be summarized into several groups of methods: those for dealing with input uncertainty; those applied to weighting and ranking; models and decision support systems (DSS) delivering highly processed and summarized derived layers into the analysis; various cognitive and soft systems methods requiring transformation for use, or perhaps sitting outside of the standard MCA; optimization approaches; and integrated spatial DSS, participatory geographical information (GIS) and multiagent systems. However, the quantification, metrication, and summary of spatial and temporal signals and temporal change in spatial patterns represent a level of sophistication and derivation that has yet to be fully explored. Recent experience with the development of simple scenario tools for assessing carbon outcomes from management change in rangelands[9,10] has emphasized the importance of spatial gradients, interactions and patterns, and temporal trends and transitions in response to anthropogenic and environmental forcing. In this chapter, the Australian rangelands are used as an example coupled human-environment system to examine the role that spatial and temporal information can play in a multicriteria framework aimed at informing policy and by definition requiring a substantial element of social context.

There is a large and long-standing literature base dealing with signal processing[11] and time series analysis[12,13,14] and merging methods across these two areas.[15] This literature indicates how the properties of demographic, economic, social, and biophysical point-based time series data can be captured. With spatially explicit time series we are interested in how these properties can be meaningfully mapped into a multicriteria analysis framework.

2.2 CONCEPT

The premise behind this chapter is that some form of multicriteria framework is useful for exploration of complex coupled human environment systems and for informing policy decision making. Integration of nonscientific knowledge is of key importance, and the user perspective may be the ultimate criterion for evaluation.[16] A requirement of this analysis is that it is simple and transparent to the client, stakeholder, participant, and decision maker, but that it has the capability to capture complex spatial and temporal interactions and trends that influence the nature of both system behavior and evolution and the consequences of decisions. In principle, it is necessary for multicriteria frameworks to include measures of system dynamics—both spatial and temporal. Therefore, the underlying theme in this chapter is the efficacy, efficiency, and information content of transformations of spatial and temporal trends, patterns, and dynamics into standardized, indexed layers for use in spatial multicriteria analysis. The ensuing discussion does not imply that multicriteria approaches are either the only way or the best way to approach analysis for policy decision making in coupled human environment systems. It is simply one approach that has proven to be useful,[4,6,17,18] and it provides a context for discussion of the issue of transformation of spatial and temporal signals out of a complex multidimensional response space into standardized, unitless, ordinal scalars to assist in human problem exploration and decision making.

2.3 TRANSFORMATION ISSUES

In terms of definition, transformation is taken to mean a method by which a more complex spatial pattern or relationship, or temporal pattern or trend, is mapped into one to many quantitative metrics that have some functional relationship or understandable descriptive contribution that can be ranked in terms of the objective of a multicriteria approach. This transformation can therefore be a simple regression function wherein the slope is used as the metric, or it can be a set of partial metrics that together provide a composite indicator capable of being ranked. Examples of the latter might include several spatial patch metrics such as number, size, and edge length or several curve metrics such as timings, amplitude, and area under the curve.

Sexton et al.[19] define four dimensions of scale:

(a) Biological—from cell, organism, population, community, ecosystem, landscape, biome to biosphere; with four useful levels: (1) genetic, (2) species, (3) ecosystem, and (4) landscape.
(b) Temporal—different spans of time for different events and processes.
(c) Social—example scheme: (1) primary interaction—physical human contact with ecosystem, (2) secondary interaction—emotional (laws, policies, regulation, votes, plans, assessments, and so forth), (3) tertiary—indirect and qualitative (values, interests, cultures, heritage, and so forth).
(d) Spatial—many hierarchies based on numerous attributes.

Possibly the greatest issue in transformation relates to scale-dependent effects. This is particularly so in human environment interactions where geographical variation in human behavior and biophysical factors at different scales interact.[20] This is also particularly so when combining biophysical data with economic and social data where pixels and polygons with discrete spatial properties must be combined with individual behaviors and institutional arrangements that operate in a multivariate pseudospatial sphere of influence[21] and have nonequivalent descriptions.[7] For example, a region may be bound by certain rules that govern the degree of economic support for certain activities. The potential spatial dimensions are the region boundary, but the effective spatial pattern inside the region is governed by a range of existing conditions, human characteristics and behaviors, economic conditions, and biophysical limitations, some of which can be directly supplied as spatial data layers, and some of which require a model of potential influence or effect to create an index of likelihood of adoption or compliance. It is possible to establish equivalence rules between biophysical and social landscape elements using structural (e.g., species composition and hydrological system versus population composition and transportation and communication infrastructure), functional (e.g., patch connectivity versus commuting), and change-based (e.g., desertification versus urbanization)[22] approaches. It is also possible to establish demographic scale equivalence between biophysical and social domains using a spatial hierarchy based on individuals (e.g., plants and people), landscapes (e.g., watersheds and counties), physiographic regions (e.g., ecoregion and census region), and extended regions (e.g., biome and continent).[22]

Relationships of information derived within one scale category are reliant on assumptions from others.[19] In a more general sense, the modifiable areal unit problem (MAUP), where correlations between layers vary with different reporting boundaries, requires excellent transformation methods, using finer scale data to inform the broader scale analysis,[23] and constant awareness of the potential problem of understanding and managing patterns, processes, relationships, and human actions at several scales.[19] Multiagent simulation approaches[24] have considerable benefits in dealing with individual behaviors in urban and densely populated system problems[25] as well as land cover change problems[26] and technology diffusion and resource use change.[27] They may also be applied to examine emergent properties at the macroscale from different microscale outcomes and incorporate spatial metrics.[28]

The second major issue in transformation relates to a meaning or quantifiable relationship with an attribute that affects or contributes to assessment of the objective of the analysis. A key element here is the fieldwork and analytical work to develop specific and general quantitative, probabilistic, or qualitative relationships between patterns and processes[29,30,31] that can be used either locally or globally to assign a rank in terms of some multicriteria objective. Laney[32] describes two approaches: studies identifying the land cover and change pattern, then seeking to develop a model to explain these patterns (pattern-led analysis) and studies that develop a theory to guide pattern characterization (process-led analysis). Both approaches may have flaws, with pattern-led analysis being highly data dependent and able to identify only processes associated with that data, and process-led analysis dependent on the prior theoretical model, adherence to which may preclude treatment of other equally valid processes and paths. The ultimate integration of transformation and meaning might be represented by the "syndrome" approach,[33] wherein alternative archetypal, dynamic, coevolution patterns of civilization-nature interaction are defined (e.g., desertification syndrome). These syndromes might be characterized by highly developed composite indicators that incorporate complex derived spatiotemporal relationships and patterns.

2.4 TRANSFORMATION DOMAINS AND METHODS

The effectiveness of a multicriteria framework is probably proportional to the extent to which system elements and interactions are captured. Representation of time in traditional GIS platforms is very poor,[34] while image-processing systems that handle time series of spatial data lack the tools for extraction and summary of information from the time domain. More accessible space-time analytical functionality is needed to make a wide variety of transformation approaches available to those other than expert spatial analysts and signal processors. The challenge lies in acquiring data in all of the potential response domains at a suitable scale and with acceptable quality. A list of possible information domains is given in Table 2.1 along with the kind of transformation issue involved and some possible methods. Where individuals are involved, demographic information coupled with surveys and units of community aggregation form the basis for transformation—spatially in terms of the location of behaviors and recorded preferences in relation to land use patterns and changes, and temporally in the sense that trajectories in opinion and behavior lead to land use change. Social systems are reflexively complex (i.e., having awareness and purpose). Therefore, within a social multicriteria analysis with nonequivalent observers and nonequivalent observations, there is a need to define importance for actors and relevance for the system.[7] The actors in social networks that influence the land use outcome must be spatially represented,[35] but there is a challenge in capturing the link between influence and biophysical outcome.[36]

At the level of social and economic statistics, collection units often determine the nature of the analysis. Social indicator data may be idiosyncratic at the local scale, have incomplete time series, have definitional changes over time, and have misaligned reporting boundaries.[37] This results in MAUP, ecological fallacy, expedient choice of statistics, arbitrary choice of measures, and difficulty in establishing any causal relationships.[37,38] Transformations are required to summarize temporal trends and cycles and to define spatial patterns and relationships at a finer scale, which may help to distribute the information downward from the collection unit in scale in a spatially explicit way. Dasymetric mapping can be used only to assign populations to remotely

TABLE 2.1
Transformation Domains for Spatiotemporal Multicriteria Frameworks

Information Domain	Transformation Issue	Methods
Individual behaviors and preferences	Representation of individual at resolution of analysis	Transform survey information into statistics and metrics that summarize the tendencies in the population for that spatial unit
Individual perceptions	Representation of abstract concepts such as beauty, degree of space contamination, etc.	Use landscape image metrics, spatial distances and landscape contents
Institutional arrangements, government regulations, and incentives	Representation of the influence or likelihood of adoption or compliance	Develop probability models based on prior surveys of impact and create probability layers
Economic variables	Relating collection unit to analysis unit	Self-organization of spatial units; temporal trends, metrics, and time period summaries
Social statistics, societal systems, transport and surveys	Conversion to a factor layer—attaching a meaning and a rank	Develop probability models and partial regression models to ascribe some of variation in target issue to the social factors. Create factor layers based on the percentage variation described, direction (+ or –) and strength (slope) of trend
Climate	Impact/response an outcome of complex temporal sequences and spatial patterns	Develop impact threshold and severity layers based on multiple scenario runs
Disturbances—fire, grazing, clearing, flooding, desertification, urbanization, abandonment	Representation of spatial extent, spatial gradient, timing, duration, impact, agents (i.e., active units such as animals)	Derive metrics describing spatial and temporal patterns, harmonics, limits, responses, demographics that can be ranked in terms of the target issue
Land use	Representation of persistence and change at level of cover type, species, management practice, seasonal magnitude	Derive metrics that capture pattern, change, persistence, sequence, and all quantitative properties of the change in a hierarchical structure
Bio/geochemical process—hydrology, sedimentation, nutrients, gas exchange, emissions, consumption	Representation of process in terms of outcome affecting or influencing target issue	Aggregated, averaged, summarized and probability converted outputs from process modeling

sensed urban classes, and population surfaces can be created by associating the count with a centroid and distributing it according to a weighted distance function.[38,39] The relationship between people and their environment is captured by cognitive appraisal from perceived environmental quality indicators.[40] Indicators of residential quality and neighborhood attachment[40] might be transformed into spatial properties by assigning proximity functions to services, assigning distance metrics to road access and access to green space, ranking buildings for aesthetics and quality of human environment, and mapping these with spatially explicit viewability constraints.

Climate provides an overarching influence that is both spatially generalized and locally spatially dependent, and it is a fundamentally time-dependent and cyclical factor. Here the transformations include spatial patterns of microclimatic variation and temporal trends in climate change, metrics of seasonal cycles and trends, or variance in extremes. The remaining information domains are the most spatially and temporally interactive, with biogeochemical processes interacting with land use type and change highly influenced by human and other disturbances. These domains require many spatial and temporal metrics as well as higher level measures of system response in the form of outputs of spatially and temporally explicit models (e.g., hydrology).

Some methods for transforming complex spatial and temporal patterns, relationships, and signals are given in Table 2.2. These are considered in terms of the general spatial context, the specific social network data where spatial and nonspatial cognitive domains mix,[40] the visual context where views and beauty perceptions intermingle with functional and locational considerations,[41] and the temporal context where methods from nonspatial time series analysis complement methods specifically developed for time series of satellite data. The spatial and temporal contexts are discussed in more detail in the following sections; however, the example landscape context used in the discussion must first be described.

2.5 EXAMPLE LANDSCAPE CONTEXT—AUSTRALIAN RANGELANDS

The Australian rangelands provide a suitable combination of spatial and temporal dynamics and dependencies for illustration of issues surrounding transformation of spatial and temporal system properties into an MCA framework. This system is characterized by a hierarchy of scales within and across which influences, effects, relationships, and functions operate. All of the scale domains of biological, temporal, social, and spatial are relevant. The system is affected by very large-scale climate and economic factors and very small-scale spatial dependencies in habitats and landscape function. The rangelands have the following characteristics:

1. Diversity in climate, soils, and vegetation types (Figure 2.1).
2. Heavily utilized by domestic livestock.
3. Substantially infested with feral animals.
4. A significant biomass and soil carbon reserve and a source of greenhouse gas emissions through annual wildfire.
5. System principally limited by water availability.
6. Spatial interactions, patterns, and gradients substantially related to landscape scale terrain–water dynamics and anthropogenic water supply (bores).

TABLE 2.2
Methods for Exploration and Transformation of Complex Spatial and Temporal Patterns, Relationships, and Signals

Spatial[23]
Convolution filtering (moving window or kernel) containing functions from simple statistics to textural indexes to complex regression to spatial autocorrelation
Distance measures in spatial neighborhoods—association of patch, gap and shape with socioeconomic change factors[29]
Cost–distance generation of user and purpose defined analysis units
Geographically weighted regression[49] to overcome nonstationarity, spatial dependencies, and nonlinear spatial distributions, allowing classification of system parameters by a learning algorithm—self-organization
Spatial/social networks[35–50]
Resilience, fast and slow adjustment, perturbation, catastrophe, turbulence, and chaos models[51]
Bioecological models—analysis of dynamic phenomena of competition-complementarity-substitution (network as a niche); social landscape analysis in landscape ecology[22]
Neural networks—not easily interpretable from economic view
Evolutionary algorithms—genetic algorithms with binary strings; evolutionary algorithms with continuous setting and floating point values
Visual[41]
Characteristic features—lower-upper feature relationships; contour block drawings; image textures; contours and horizon; spatial relations of spaces and elements; proportions of landscape zones in view; hierarchical properties; typology of fringes
Spatial distance measures—view texture; intrusion into skyline and landscape line; relative structural complexity; relative proportions; distance–size relationships
Sensitivity—functional distance in landscapes; structural distances to be kept free
Temporal
Traditional time series analysis[12,14,52,53]—trends, cycles, seasonality, lags, phase, irregularity, smoothing, differencing, autocorrelations, spectral analysis
Curve metrics[43,54]—limits, amplitude, periodicity, timings, areas, slopes, trajectories[55]; phenology
Signal processing—Fourier transforms[56]; Wavelet transforms[15,44]
Principal component analysis of time series[44]
Complex bio-socioeconomic cycles (e.g., Kondratieff waves[57]); syndromes of change[33]

7. Temporal dynamics heavily influenced by interaction between climate (water supply), grazing and fire.
8. Meso-scale landscape properties strongly linked to overall landscape function, particularly in relation to water harvesting and consequent habitat development.
9. Significant social issues through indigenous rights and sacred sites and site-based tourism.
10. Management of the landscape is influenced by exogenous temporal variation in cost of finance and inputs, trade barriers and restrictions, price of commodities, specifically beef cattle, and changes in family structures and rural employment.

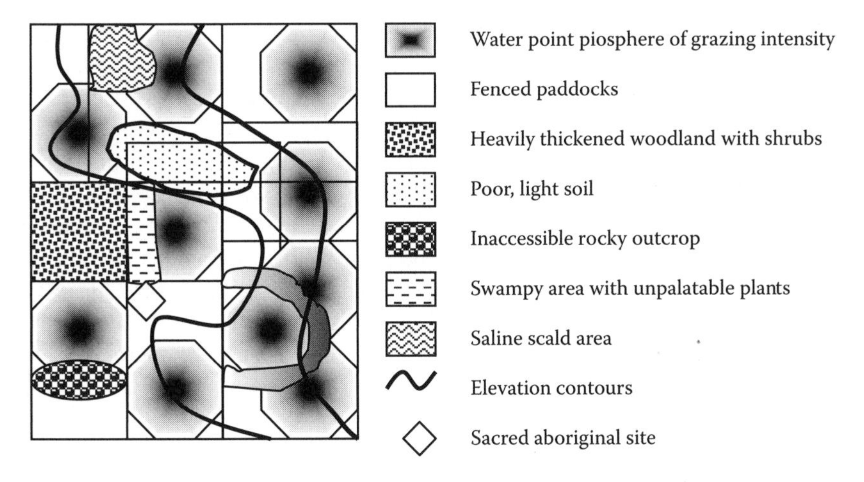

FIGURE 2.1 The concept of grazing piospheres interacting with landscape structure to create spatially and temporally dependent response zones in Australian rangelands. These are more prevalent where rainfall is less reliable, paddocks are smaller, and stocking pressure is higher.

The system represents a type of example where human demographics are not a major factor since large pastoral leases are essentially unpopulated except for the station homestead and associated buildings. Human influence in this environment is provided through management, which reaches out from the homestead to influence very large tracts of land. Hence, superficially it might be difficult to draw methodological parallels with the many coupled human environment systems worldwide and high human population densities. However, in this system, demographics are still important since the major influential population is that of domesticated beef cattle, with ancillary influence from feral animal populations. They are individual economic units with costs associated with parasite and disease control and human handling and value in terms of food and breeding potential. The decision-making framework for cattle is much less complex than for humans; cattle require water, feed, shade, and socialization and will optimize their behavior within this response space. Nevertheless, they influence and respond to spatial and temporal patterns, and, therefore, this system can still provide useful methodological insights.

2.5.1 SPATIAL PATTERNS AND RELATIONSHIPS

The spatial interrelationships in this rangeland system can be illustrated by a stylized landscape containing artificial water points surrounded by piospheres of influence by grazing animals upon the vegetation up to a distance limit (Figure 2.1). These water points occur within fenced paddocks, parts of which are inaccessible to stock since they are outside the water access limit. The paddocks also contain different land cover types with different habitat suitability, fire susceptibility, and livestock carrying capacities. The landscape has rocky areas, areas with thick shrubland inaccessible to stock, swampy and saline areas with low productivity, and an aboriginal sacred site. The area also has an aesthetic component with a viewpoint and rest area located on a major road, with basic picnic facilities outside the mapped extent. The major spatial gradients in this landscape are created by the effect of grazing on vegetation and habitat, the connectivity between habitats, the structure in relation to shelter, water harvest and stock access, and the appearance of the landscape from a specific direction and angle of view.

In order to capture spatial attributes, a level of spatial pattern reporting must be defined, and this level of aggregation must be compatible with the resolution of other data in the analysis. The scale of aggregation might relate to some functional distance and sphere of influence in the landscape, and pattern extraction might be undertaken for a number of different aggregation units,[42] a nested set of patch scales,[30] in order to specifically capture the influence of landscape structure from different elements of the system such as bird habitat, cattle grazing behavior, scale of microtopography, and so forth.

2.5.2 TEMPORAL PATTERNS AND INFLUENCES

The temporal behaviors of, and influences on, this rangeland system could be described by a time series of weather and satellite data, which records sequences of detectable land cover change and vegetation state, as well as derived measures of system function integrated through models. A monthly time series of net ecosystem carbon exchange (Barrett, personal communication) provides an example data set for illustration of approaches to disaggregation and decomposition of signals into meaningful indexes (Figure 2.2). A series of seasonally based system responses provide the basis for extraction of:

1. Curve metrics that describe the timing, duration, magnitude and periodicity of the response[43]
2. A cumulative aggregate of the net system behavior through time
3. Trend in signal from wavelet or other transforms[15,44]
4. Temporal autocorrelation to see how strong the "memory" is in the system—a strong memory indicates more regular cyclical behavior
5. Power spectrum and Fourier transforms on original data and first differences or first derivatives to detect major cyclical patterns—in this case occurring at about 22, 44, and 66 months
6. Cumulative probability curves to identify the relative behavior for some proportion of cases (Figure 2.2)

These metrics and measures of time series attributes can be derived spatially and converted to single or partial component indicators of system properties.

The temporal influences are also represented by nonbiophysical time series such as livestock numbers, climate cycle indexes, prices and costs, and human activity measures (Figure 2.3). These data may only be available at a coarse level of spatial resolution, such as cattle numbers from the agricultural census, or individual behaviors from social surveys with limited samples. Alternatively, they may be global variables such as cattle prices, interest rates, and climate indexes such as the southern oscillation index (SOI). In these cases, a means must be found to apply these spatially via some filtering layer that assigns the attributes only to those pixels where the influence occurs, or to those pixels not constrained by other factors.

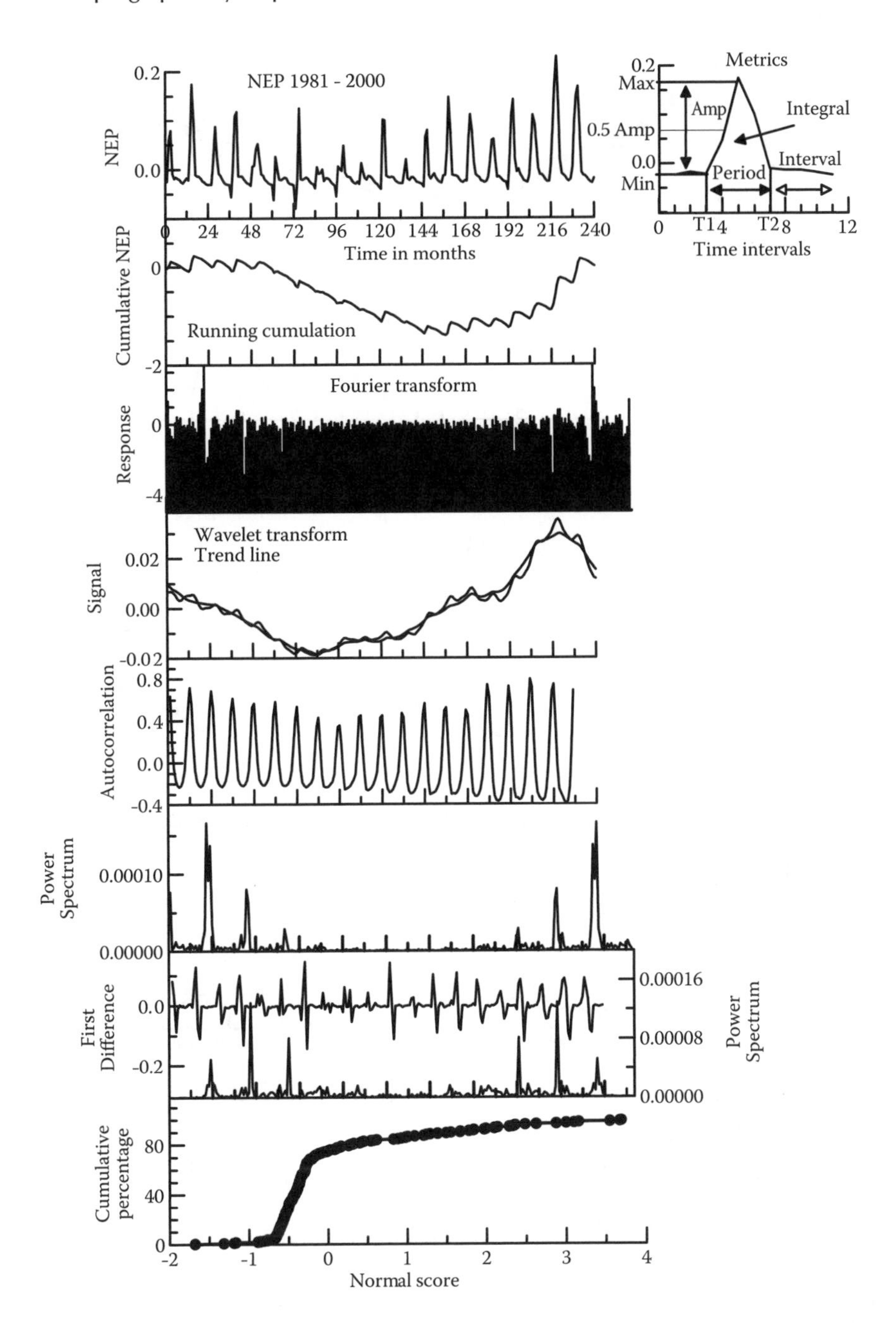

FIGURE 2.2 Time series approaches to extracting signal summary indicators. A time series of net ecosystem productivity indicates the base potential for carbon fixation. This may be transformed into indicators by calculating metrics, including a running integral, extracting the frequency of cyclic patterns using fast Fourier transform (FFT) or power spectrums on original data or first differences and derivatives, defining direction of temporal change through trend analysis or wavelet transforms, and estimating likelihood of various levels though cumulative probability.

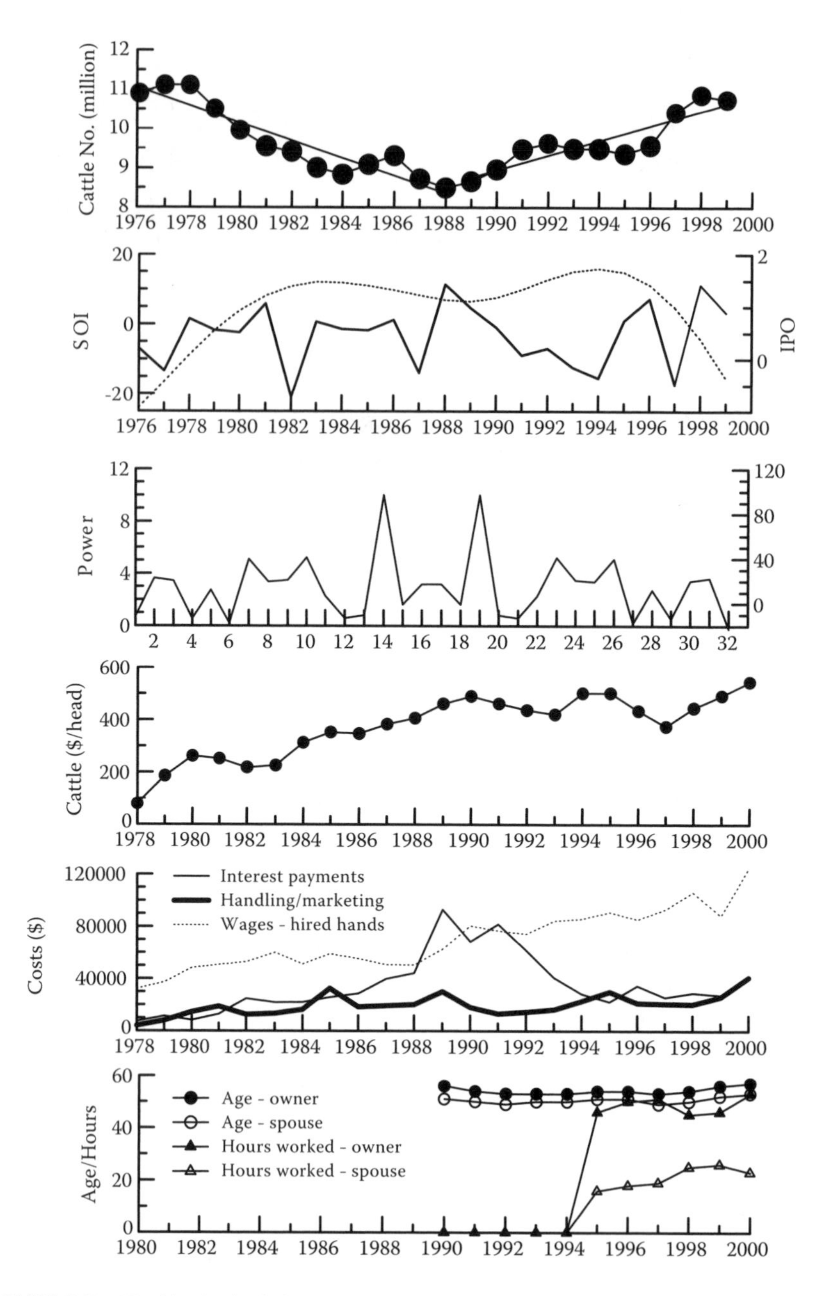

FIGURE 2.3 Nonbiophysical time series also provide potential indicators but may not be spatially explicit at the required scale. These need to be transformed into indicators such as trends in demographics (e.g., cattle), patterns of climate and frequency of occurrence of certain climate types, trends in prices for commodities, trends in costs of production, and trends in human activities potentially affecting management and economic outcomes.

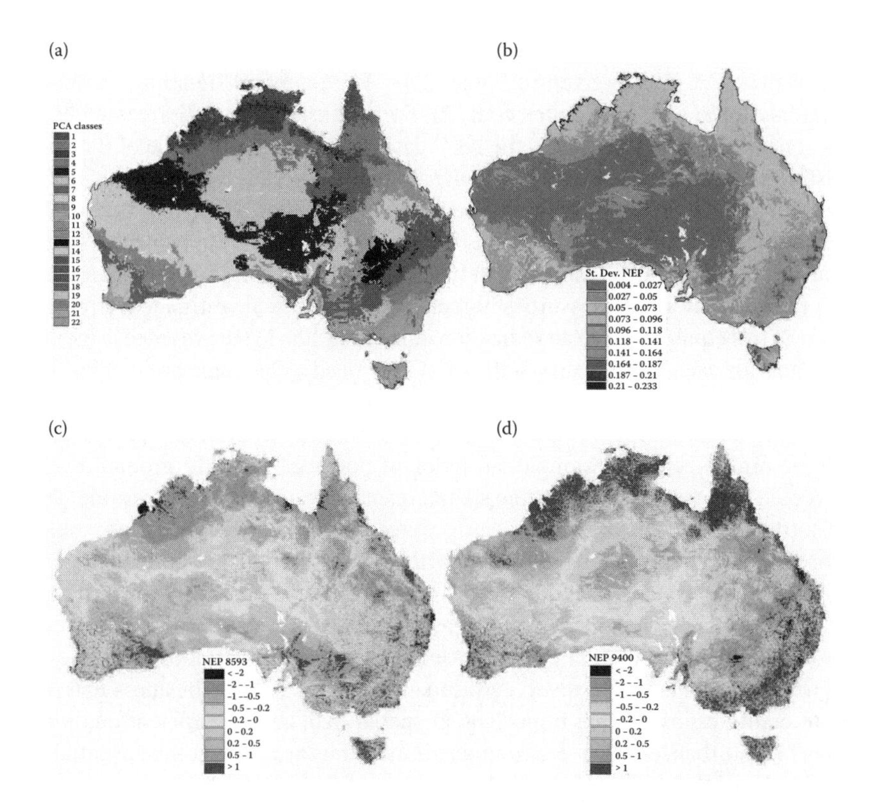

FIGURE 2.4 **(See color insert following p. 132.)** (a) The spatial pattern of temporal signals may be grouped by applying principal components analysis to the time series and creating a classification based on the major principal components. (b) The temporal metrics may be calculated on the spatial times series to create maps of, for example, standard deviation of net ecosystem productivity (NEP). (c) A running integration, trend, or wavelet analysis may define periods of distinct behavior in the time series that can then be summarized by metrics such as an integral of NEP for periods of decline and increase. Shown for 1985–1993 and 1994–2000 here.

The spatial extent of discrete and consistent temporal patterns may be defined by, for example, a principle components analysis (PCA) on the time series and subsequent classification (Figure 2.4a). This reveals distinct temporal patterns that represent a regional summary (Figure 2.4b; the temporal net ecosystem productivity (NEP) signal for one of these classes is used in Figure 2.2). By contrast, time series attributes may be calculated on a pixel-by-pixel basis, and the data, such as standard deviation in NEP, are mapped to provide a continuous factor layer. The analysis of the trends in the time series (Figure 2.2) may provide information about major periods of differing behaviors, such as the net loss carbon throughout northern Australia between 1985 and 1993, versus the net gain in carbon between 1994 and 2000 (Figure 4.2c).

2.5.3 Data and Information: Scale of Representation

Moving down scale to the region of interest for analysis, the Victoria River District (VRD) in the northern Australian rangelands provides a good basis for assessment of data scales and transformation through spatial filtering (Figure 2.5). The region has three of the PCA classes given in Figure 2.4a. The temporal signal and associated metrics and time series attributes could be broadly assigned to all areas within the class zone in the VRD. However, the NEP data represent the response of the system undisturbed by livestock. Therefore, in the absence of specific estimates that take the spatially variable impact of grazing intensity around water points into account, one could apply an arbitrary scaling of effect on NEP that varies from using the supplied signals for ungrazed areas, to completely suppressing the accumulation signal at the water point, where stock pressure is highest. Very broad scale estimates of profit per hectare at full equity provide an indication of profitability. Mine presence is indicated by a count for each 1 km pixel—a decision may need to be made about a buffering rule for radius of disturbance. The number of threatened birds is derived from a very coarse resolution data set. However, if any information is available on the habitat for these birds, then, for example, an index of potential threat to ground dwelling birds could be created with very fine spatial resolution using a rule governing degree of disturbance with distance from cattle watering points. Completely aspatial, but important, system attributes and metrics may be downscaled to appropriate resolution if relationships to spatial data at an appropriate resolution are known or can be derived. The major challenge arises in ascribing causal relationships and areas of interest to human population centers based on social statistics about activities and preferences of humans. However, certain key variables such as business enterprise debt to equity ratios may be important. If spatial data are of sufficient quality and effects of disturbance on key environmental measures are known, then aspatial economic and social data may be used along with biophysical data to ascribe integrative indexes of economic benefit, cost of degradation, and social benefit to individual landscape pixels having particular suites of biophysical attributes.

2.5.4 Some Spatiotemporal Inputs to a Rangeland MCA

The rangeland example provides a very specific opportunity to elucidate the etiology of spatial and temporal transformations to summarize complex system responses and behaviors in multicriteria analysis. If we take the effect of grazing on vegetation condition as an example of a complex biophysical process influenced by human management, in order to capture the elements of the condition of a particular pixel one might need:

1. Distance to water point (see Figure 2.1)—simple distance metric.
2. Value of average grazing pressure—complex functional calculation based on distance functions describing the cost–distance relationship between distance from water point and livestock tendency to travel distance from water. Then the value of average grazing pressure is calculated from the ratio of average density of livestock units to the nominal safe carrying capacity of the vegetation type. This is related to time-use analysis,[45] where the period of time that an individual unit is in a particular spatial location is important.
3. A proximity function to dense shrubby vegetation—modifies (2) above, since grazing pattern may be perturbed by animal behavior with respect to shelter.
4. Average maximum seasonal biomass for that pixel—simple temporal curve metric.

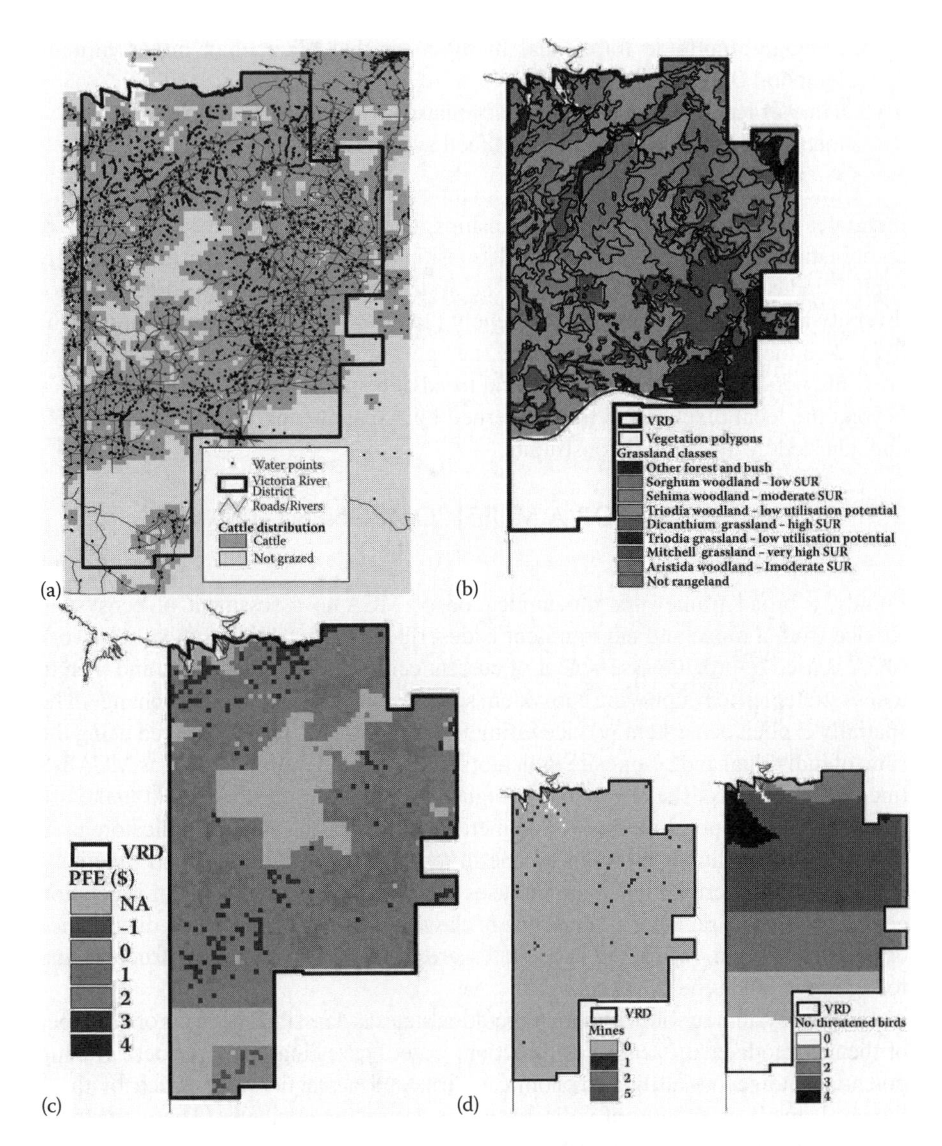

FIGURE 2.5 **(See color insert following p. 132.)** Temporal signals are usually based on biophysical or human phenomena that operate at a large scale (e.g., climate, interest rates). Demographic changes at a fine scale may have scale limitations due to level of aggregation in reporting. Temporal signals and indicators are filtered by spatial variation. The Victoria River District in the Northern Territory of Australia is highly productive. (a) Cattle are distributed of freehold-leasehold land but confined by water points. (b) Both productivity and ecological impact vary with vegetation type, which is associated with soils, topography, and rainfall gradient. (c) Costs are low and enterprises are profitable but the increment is small on a per hectare basis. (d) Mining with major physical disturbance occurs sporadically across the area. There are threatened bird species in the region and these may be ground nesting and impacted by grazing.

5. Average amplitude for annual biomass for that pixel (max minus min) temporal curve metric.
6. Time of half maximum seasonal biomass for that pixel—simple temporal metric for start of period of green feed availability.

These data might be used to define an index of animal impact for each pixel that integrates spatial and temporal relationships, trends, and influences. This kind of combination of spatial and temporal relationships may be used to construct spatially explicit indexes for other elements of the system such as landscape function, biodiversity impact, and socioeconomic benefit (Table 2.3). The temporal component is filtered on the basis of spatial relevance (i.e., grazed areas are relevant but ungrazed areas are not), except where the temporal trend of response has a sphere of influence beyond the local pixel and is then governed by a spatial function for relative effect and adjusted by any spatial constraints.

2.6 A FRAMEWORK FOR A MULTICOMPONENT ANALYSIS WITH MCA

Finally, a broad framework for application of MCA to assessment of ecosystem service from a rangeland environment is described (Figure 2.6). In this framework MCA is used to provide assessment of current ecosystem service levels, and then to assess strategies for improving ecosystem services through management change. The spatially explicit ecosystem service rating for an area could be constructed using the suite of individual and composite indicators in a MCA environment such as MCAS-S (multicriteria analysis shell–spatial[46]; Figure 2.7). This framework could make use of spatial and temporal measures and metrics as part of the suite of indicators used to define the condition of the landscape in terms of a range of uses and functions. The approach described in Figure 2.6 uses state and transition models of vegetation condition.[10,11] The rangeland landscape is classified into states based on disturbance of original natural vegetation. The states are described in terms of structure and foliage cover and type of vegetation.[47]

The ecosystem service from the vegetation states is described in terms of a number of themes: biodiversity, landscape function, water harvesting, carbon stock, grazing potential, indigenous utility, economic return, and aesthetic value. Each of these themes is made up of a set of indicator layers describing attributes. These attributes could be individual measures such as number of threatened birds, carbon biomass, or proximity to aboriginal sacred sites. Alternatively they could be composite indicators based on aggregation of individual measures in complex spatiotemporal relationships to give animal impact per pixel. These individual and composite indicators may be combined by various methods into a single index of ecosystem service for that theme. Overall ecosystem service from that landscape pixel is represented by the combination of the individual theme indicators into one overall index. The overall ecosystem services level can be improved by moving the theme indicators to higher levels. Each theme can be improved by applying a number of strategies—individual strategies may improve more than one theme, but may have a negative effect in other themes. For example, economic potential may be increased by increasing stocking rates, but this may affect various elements of landscape function.

TABLE 2.3
Derived Factor Layers That Integrate Spatial and Temporal Trends with Function and Ecosystem Outcome

Factor	Input data	Spatial	Input data	Temporal	Integration
Animal impact per pixel	Buffered water points with animal distribution assigned by exponential distance function; map of inaccessible or unattractive vegetation	Probability of animal utilization in any time period—from distance function of frequency of visitation modified by spatial accessibility and divided by feed requirement for specified weight gain	Modeled time series of grassland growth; estimate of feed requirement per unit of time; estimate of safe utilization rate for vegetation types	Likelihood that animal demand will not exceed safe utilization rate in any period—from historical time series of feed accumulation in growing seasons	Combined, scaled index of relative animal impact on any piosphere pixel due to spatial control by water access and temporal probability of overgrazing
Landscape function	Elevation, slope, high resolution imagery defining trees and bushes; derived landscape drainage pattern and water accumulation	Textural analysis of pattern of water harvest converted to magnitude and density measures	Time series of rainfall and evaporation; modeled demand for water by vegetation types	Timing, frequency, and periodicity of water flows and probability of water harvest sufficient to sustain stable system	Combined, scaled index of water harvest function for any pixel due to spatial pattern of landscape elements and temporal probability of successful capture if harvestable rainfall events occur
Socioeconomic benefit per grazed unit	Live weight gain per animal; animal distribution; value of live weight (price per animal); cost of production (cost per animal)	Percentage of total live weight gain attributable to each grazed pixel then cost, price and profit per grazed pixel; spatial correlation with vegetation type.	Time series of animal prices, cost changes, animal numbers (stocking rates), feed availability, live weight gain per unit time.	Length of periods of increase or decline in terms of trade and in profit per unit of grazed area	Combined, scaled index of economic return per unit area due to spatial distribution of animals and feed, and temporal changes in prices and costs per unit area
Biodiversity impact per pixel	Distribution of endangered species; association of species with vegetation, landscape features	Connectivity, pattern of habitat; focus and extent of disturbance from animal impact per pixel	Time series of species numbers; climate, feed supplies, stock numbers, predator numbers	Periodicity of population fluctuations; timing of lows in population; trends in population	Combined, scaled index of probability of negative biodiversity impact from reduced connectivity and disturbance during population lows

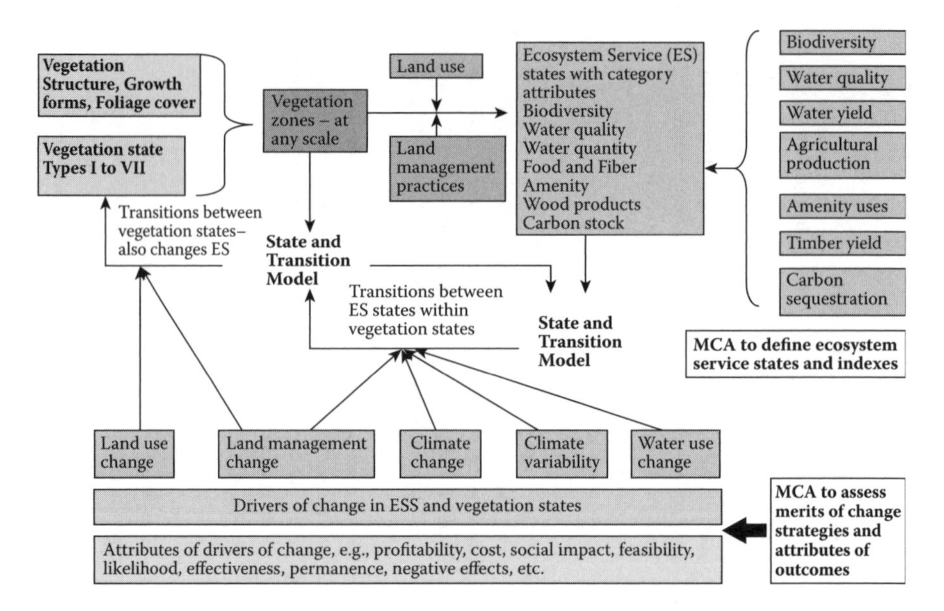

FIGURE 2.6 A framework for application of multicriteria analysis of ecosystem services to the rangeland environment.

Strategies for improving ecosystem service within a land cover state may be assessed by a mixture of spatial and nonspatial MCA where drivers of changes within each state are defined, and feasibility or effectiveness of adjustment to these drivers is assessed using a range of spatial attributes. An example of the combination of a variety of data layers by a relatively simple method to form a single index of potential productivity from grazing in Australian rangelands is given in Figure 2.7. This could represent a single theme within a full ecosystem services' assessment. These themes may be compared by two-way analysis[18,46] or may be combined and an overall condition assessed using spider diagrams/radar plots. Some detailed contextual analysis of this kind of ecosystem service assessment incorporating the use of spatial data and spider plots is provided by Defries et al.[48]

2.7 CONCLUSIONS

Transformation of complex spatial and temporal patterns and behaviors into simple data forms is an important enabling technical capability required in order to maximize the information content and effectiveness of digital decision support systems. An essential framework of analytical capability involves seamless access to tools and methods for spatial analysis and extraction of spatial patterns and interrelationships, tools for time series analysis on spatial data, tools for developing simple models and defining relationships and causality across spatial scales, and generation of uncertainty and error measures and incorporation in analysis along with base data. A schema for transformation of spatiotemporal complexity into indicators for MCA is given in Figure 2.8. The schema uses rangeland and urban examples to illustrate the kinds of specific measures and derived information required. Process

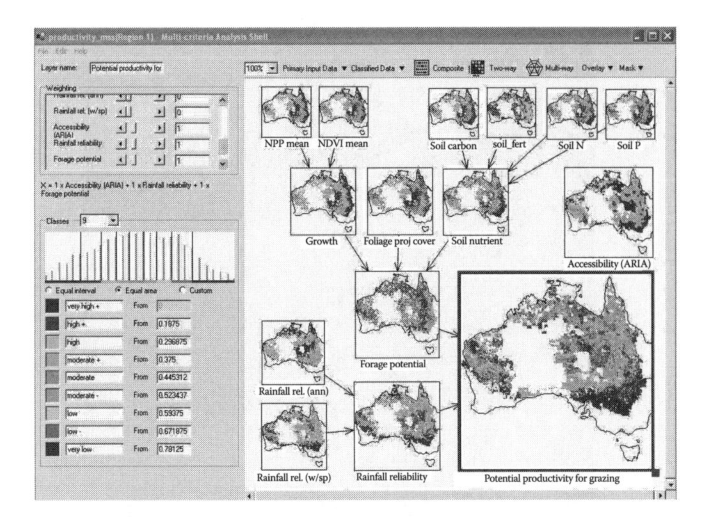

FIGURE 2.7 **(See color insert following p. 132.)** Cognitive mapping interface suitable for combination of diverse spatiotemporal metrics, indices, and data layers describing meaningful properties of a system under analysis. This example shows the construction of a composite index to represent potential grazing productivity from rangelands.[46,58]

Conceptual Basis for Transformation

Temporal variation, e.g., climate; economic conditions

Variation in available resources

Space with gradients and objects
e.g., landscape, city

function, appearance

density, range

Mobile agent making decisions
e.g., cattle/humans

Spatial relationships and patterns
e.g., terrain/soil; buildings/open space

texture, association

quantity, type

Spatial relationships and patterns
e.g., feed and water requirement; play areas

Spatial constraints, e.g., distance from water or business clients and services
Temporal constraints, e.g., time spent in location or decision/consulting time

Measures of system properties and processes – Indicators

Temporal dependencies
variation in feed supply
probability of demand > supply
frequency of drought
travel time
recreation time

Interrelation and combination

Spatial impacts
grazing/utilization pressure
trampling/tracking
pre-condition for woody ingress ease of business interaction attractiveness for tourists

FIGURE 2.8 Schema for transformation of spatiotemporal information into indicators of system properties for multicriteria analysis evaluations.

FIGURE 2.9 Vegetation types in Australian rangelands. (a) Northern savanna woodlands may have excellent herbaceous cover due to large paddocks, lower stock density and reliable seasonal rainfall (Photo M. J. Hill). (b) Saltbush plains may be in good condition (as shown here) but can be susceptible to degradation with droughts and overgrazing (Photo courtesy of R. Lesslie).

and other complex models operate externally to this framework and provide spatial layers and temporal signals representing an integration of complex processes for use within the framework.

The example used in this chapter largely addresses the biophysical domain since rangelands have low human populations and individuals have control over large land areas. This may mean that sociological factors play a smaller relative role in the management of the system except when concerned with indigenous rights and issues. However, the analytical problems are universal—for example, there is still a need to determine the level and importance of changes in spousal work contributions to the operation of cattle stations and to the functioning of isolated families even if these effects are very small. However, through the principle of equivalence of biophysical and social landscape elements, cattle demographics, like human demographics in cities, operate as a major driver in the system, and assigning rules to behavior and defining spatial relationships and temporal trends and influences assume critical importance. This chapter has provided some examples of analytical methods and approaches needed. The case studies to follow will examine spatial methods in greater detail for a range of geographically distinct systems and problems.

2.8 ACKNOWLEDGMENTS

The MCAS software interface illustrated in this chapter (Figure 2.7) has been developed by Rob Lesslie as team leader, Andrew Barry as programmer, and the author as technical advisor. I am grateful to my colleagues for permission to reproduce work from our joint efforts in this sole author chapter. I thank Damian Barrett for access to continental carbon cycle model outputs used in Figures 2.2 and 2.4.

REFERENCES

1. Grant, W. E., Peterson, T. R., and Peterson, M. J. Quantitative modelling of coupled natural/human systems: simulation of societal constraints on environmental action drawing on Luhmann's social theory. *Ecological Modelling* 158, 143–165, 2002.
2. Jankowski, P. Integrating geographical information systems and multi-criteria decision making methods. *International Journal of Geographical Information Systems* 9, 251–273, 1995.
3. Varma, V. K., Ferguson, I., and Wild, I. Decision support system for the sustainable forest management. *Forest Ecology and Management* 128, 49–55, 2000.
4. Store, R., and Jokimaki, J. A GIS-based multi-scale approach to habitat suitability modelling. *Ecological Modelling* 169, 1–15, 2003.
5. Mazzetto, F., and Bonera, R. MEACROS: A tool for multi-criteria evaluation of alternative cropping systems. *European Journal of Agronomy* 18, 379–387, 2003.
6. Giupponi, C. et al. MULINO-DSS: A computer tool for sustainable use of water resources at the catchment scale. *Mathematics and Computers in Simulation* 64, 13–24, 2004.
7. Munda, G. Social multi-criteria evaluation for urban sustainability policies. *Land Use Policy* 23, 86–94, 2006.
8. Hill, M. J. et al. Multi-criteria decision analysis in spatial decision support: The ASSESS analytic hierarchy process and the role of quantitative methods and spatially explicit analysis. *Environmental Modelling and Software* 20, 955–976, 2005.
9. Hill, M. J. et al. Vegetation state change and consequent carbon dynamics in savanna woodlands of Australia in response to grazing, drought and fire: A scenario approach using 113 years of synthetic annual fire and grassland growth. *Australian Journal of Botany* 53, 715–739, 2005.
10. Hill, M. J. et al. Analysis of soil carbon outcomes from interaction between climate and grazing pressure in Australian rangelands using Range-ASSESS. *Environmental Modelling and Software* 21, 779–801, 2006.
11. Smith, S. W. *The Scientist and Engineer's Guide to Digital Signal Processing,* 2nd ed. California Technical Publishing, San Diego, Calif. 643 pp, 1999.
12. Hamilton, J. D. *Time Series Analysis.* Princeton University Press, N.J. 820 pp, 1994.
13. Gourieroux, C., and Monford, A. *Time Series and Dynamic Models.* Cambridge University Press, UK. 668 pp, 2000.
14. Brockwell, P., and Davis, R. *Introduction to Time Series and Forecasting.* Springer Verlag, Netherlands. 434 pp, 2002.
15. Percival, D. B., and Walden, A. T. *Wavelet Methods for Time Series Analysis (WMTSA).* Cambridge University Press, Cambridge. 594 pp, 2000.

16. Siebenhuner, B., and Barth, V. The role of computer modelling in participatory integrated assessments. *Environmental Impact Assessment Review* 25, 391–409, 2004.
17. Walker, J. et al. *Assessment of Catchment Condition*. CSIRO Land and Water, Canberra. 36 pp, 2002.
18. Hill, M. J. et al. Multi-criteria assessment of tensions in resource use at continental scale: a proof of concept with Australian rangelands. *Environmental Management* 37, 712–731, 2006.
19. Sexton, W. T., Dull, C. W., and Szaro, R. C. Implementing ecosystem management: A framework for remotely sensed information at multiple scales. *Landscape and Urban Planning* 40, 173–184, 1998.
20. Osborne, P. E., and Suarez-Seoane, S. Should data be partitioned spatially before building large-scale distribution models? *Ecological Modelling* 157, 249–259, 2002.
21. Geoghegan, J. et al. Modelling tropical deforestation in the southern Yucatan peninsular region: comparing survey and satellite data. *Agriculture, Ecosystems and Environment* 85, 25–46, 2001.
22. Field, D. R. et al. Reaffirming social landscape analysis in landscape ecology: a conceptual framework. *Society and Natural Resources* 16, 349–361, 2003.
23. Nelson, A. Analysing data across geographic scales in Honduras: Detecting levels of organisation within systems. *Agriculture, Ecosystems and Environment* 85, 107–131, 2001.
24. Bousquet, F., and Le Page, C. Multi-agent simulations and ecosystem management: A review. *Ecological Modelling* 176, 313–332, 2004.
25. Loibl, W., and Toetzer, T. Modeling growth and densification processes in suburban regions—simulation of landscape transition with spatial agents. *Environmental Modelling and Software* 18, 553–563, 2003.
26. Evans, T. P., and Kelley, H. Multi-scale analysis of a household level agent-based model of landcover change. *Journal of Environmental Management* 72, 57–72, 2004.
27. Berger, T. Agent-based spatial models applied to agriculture: A simulation tool for technology diffusion, resource use changes and policy analysis. *Agricultural Economics* 25, 245–260, 2001.
28. Parker, D. C., and Meretsky, V. Measuring pattern outcomes in an agent-based model of edge-effect externalities using spatial metrics. *Agriculture, Ecosystems and Environment* 101, 233–250, 2004.
29. Peralta, P., and Mather, P. An analysis of deforestation patterns in the extractive reserves of Acre, Amazonia from satellite imagery: a landscape ecological approach. *International Journal of Remote Sensing* 21, 2555–2570, 2000.
30. Crews-Meyer, K. A. Agricultural landscape change and stability in northeast Thailand: Historical patch-level analysis. *Agriculture, Ecosystems and Environment* 101, 155–169, 2004.
31. Laney, R. A process-led approach to modelling land use change in agricultural landscapes: A case study from Madagascar. *Agriculture, Ecosystems and Environment* 101, 135–153, 2004.
32. McConnell, W. J., Sweeney, S. P., and Mulley, B. Physical and social access to land: Spatio-temporal patterns of agricultural expansion in Madagascar. *Agriculture, Ecosystems and Environment* 101, 171–184, 2004.
33. Cassel-Gintz, M., and Petschel-Held, G. GIS-based assessment of the threat to world forests by patterns on non-sustainable civilisation-nature interaction. *Journal of Environmental Management* 59, 279–298, 2000.
34. Sui, D. Z. GIS-based urban modelling: Practices, problems and prospects. *International Journal of Geographical Information Science* 12, 651–671, 1998.
35. Faust, K. et al. Spatial arrangements of social and economic networks among villages in Nang Rong District, Thailand. *Social Networks* 21, 311–337, 1999.

36. Leenders, R. Th. A. J. Modeling social influence through network autocorrelation: Constructing the weight matrix. *Social Networks* 24, 21–47, 2002.
37. Jackson, J. E., Lee, R. G., and Sommers, P. Monitoring the community impacts of the Northwest Forest Plan: An alternative to social indicators. *Society and Natural Resources* 17, 223–233, 2004.
38. Martin, D., and Bracken, I. The integration of socio-economic and physical resource data for applied land management information systems. *Applied Geography* 13, 45–53, 1993.
39. Martin, D. Automatic neighbourhood identification from population surfaces. *Computers, Environment and Urban Systems* 22, 107–120, 1998.
40. Bonaiuto, M., Fornara, F., and Bonnes, M. Indexes of perceived residential environment quality and neighbourhood attachment in urban environments: A confirmation study on the city of Rome. *Landscape and Urban Planning* 65, 41–52, 2003.
41. Krause, C. L. Our visual landscape: managing the landscape under special consideration of visual aspects. *Landscape and Urban Planning* 54, 239–254, 2001.
42. Croissant, C. Landscape patterns and parcel boundaries: An analysis of composition and configuration of land use and land cover in south-central Indiana. *Agriculture, Ecosystems and Environment* 101, 219–232, 2004.
43. Reed, B. C. et al. Measuring phenological variability from satellite imagery. *Journal of Vegetation Science* 5, 703–714, 1994.
44. Li, Z., and Kafatos, M. Interannual variability of vegetation in the United States and its relation to El Nino/Southern Oscillation. *Remote Sensing of Environment* 71, 239–247, 2000.
45. Zhang, M. Exploring the relationship between urban form and nonwork travel through time use analysis. *Landscape and Urban Planning* 73, 244–261, 2005.
46. Hill, M. J. et al. A simple, portable, multi criteria analysis tool for natural resource management assessments. In: Zerger, A, and Argent, R. M., eds., *MODSIM 2005, Proceedings of the International Symposium on Modelling and Simulation.* University of Melbourne, 12–15 December, 2005, CDROM (available at: http://www.mssanz.org.au/modsim05/authorsH-K.htm#h).
47. Thackway, R., and Lesslie, R. J. Reporting vegetation condition using the Vegetation Assets, States and Transitions (VAST) framework. *Ecological Management and Restoration* 7, s53–s61, doi: 10.1111/j1442-8903.2006.00292.x., 2006.
48. Defries, R. S., Asner, G. P., and Houghton, R. Trade-offs in land use decisions: Towards a framework for assessing multiple ecosystem responses to land-use change. In: Defries, R. S., Houghton, R., and Asner, G. P., eds. *Ecosystems and Land Use Change.* Geophysical Monograph Series 153, American Geophysical Union, 2004.
49. Platt, R. V. Global and local analysis of fragmentation in a mountain region of Colorado. *Agriculture, Ecosystems and Environment* 101, 207–218, 2004.
50. Reggiani, A., Nijkamp, P., and Sabella, E. New advances in spatial network modelling: Towards evolutionary algorithms. *European Journal of Operational Research* 128, 385–401, 2001.
51. Nijkamp, P., and Reggiani, A. Non-linear evolution of dynamic spatial systems. The relevance of chaos and ecologically-based models. *Regional Science and Urban Economics* 25, 183–210, 1995.
52. Roderick, M. L., Noble, I. R., and Cridland, S. W. Estimating woody and herbaceous vegetation cover from time series satellite observations. *Global Ecology and Biogeography 8,* 501–508, 1999.
53. Lu, H. et al. Decomposition of vegetation cover into woody and herbaceous components using AVHRR NDVI time series. *Remote Sensing of Environment* 86, 1–18, 2003.
54. Malingreau, J.-P. Global vegetation dynamics: satellite observations over Asia. *International Journal of Remote Sensing* 9, 1121–1146, 1986.

55. Zhang, X. et al. Monitoring vegetation phenology using MODIS. *Remote Sensing of Environment* 84, 471–475, 2003.
56. Moody, A., and Johnson, D. M. Land-surface phenologies from AVHRR using the discrete fourier transform. *Remote Sensing of Environment* 75, 305–323, 2001.
57. Devezas, T., and Corredine, J. T. The biological determinants of long wave behavior in socio-economic growth and development. *Technological Forecasting and Social Change* 68, 1–57, 2001.
58. Lesslie, R. et al. *Towards Sustainability for Australia's Rangelands. Analysing the Options.* Department of Agriculture, Fisheries and Forestry, Australia, 16 pp, 2006.

Part II

Comparative Regional Case Studies

3 Spatial Methodologies for Integrating Social and Biophysical Data at a Regional or Catchment Scale

Ian Byron and Robert Lesslie

CONTENTS

3.1 INTRODUCTION

Catchment or watershed-based approaches to natural resource management and planning have been widely adopted in many countries across the globe including Australia.[1,2,3] These approaches seek to meld the benefits of building local community engagement with the need to develop better integrated and coordinated approaches for addressing landscape-scale changes in the condition of land and water resources.[4]

In Australia, regional catchment management organizations now manage a large proportion of national investment in natural resource management through the Natural Heritage Trust and the National Action Plan for Salinity and Water Quality. The Natural Heritage Trust represents Australia's largest investment in environmental management.[5] As part of the delivery of funds, catchment groups are required to develop regional plans that set out how the natural resources of the region are to be managed. Each regional plan is to be endorsed by state and Australian government agencies prior to their implementation. Although there are state and regional differences, these catchment groups are typically asked to:

- Describe their catchment condition in terms of environmental, economic, and social assets
- Identify the desired future condition of those assets
- Identify the key processes that might mitigate the achievement of the desired conditions
- Identify management actions and targets that will help achieve desired conditions
- Monitor and evaluate progress

Clearly these roles require catchment groups to be able to understand the drivers and barriers affecting land managers and understand the impacts of land management practices on key regional assets. Unfortunately, there are very limited data available that have been designed with these purposes in mind. Although most regional groups in Australia have access to a range of biophysical data sources, very few have access to detailed social data specific to their region. Endter-Wada et al.[6] asserted that natural scientists have been reluctant to include social science dimensions in ecosystem assessments. At the same time, Brown[7] suggested that in part the lack of social data incorporated in landscape planning reflects an absence of systematic approaches for collecting and analyzing this information with biophysical data.

Nevertheless, there is increasing recognition of the need to integrate social and biophysical data to achieve improved natural resource management outcomes. A review conducted as part of Australia's National Land and Water Resources Audit concluded that there is a strong need for approaches that integrate socioeconomic and biophysical data at a regional scale.[8] Similarly, Endter-Wada et al.[6] concluded that effective ecosystem management is predicated on bringing together scientific analysis of social factors and biophysical factors.

The relatively recent emergence of geographical information systems (GIS) has provided an important set of tools to facilitate interdisciplinary research that integrates social and biophysical data at a landscape scale.[9] In recent years there have been a number of important studies using census data to link changes in social structure with ecological factors and visa versa.[9,10,11] Although these studies have clearly highlighted how integrating social and biophysical data is critical to improving natural resource management outcomes, they also acknowledge significant limitations with nationally collected census data. In particular, census data are only available at aggregate levels that require researchers to assume that the social variables are homogenous across the smallest census unit (typically 200 households).[9]

In addition to concerns about spatial resolution, Endter-Wada et al.[6] suggest that while important, understanding demographic trends alone is insufficient for understanding complex social systems and their relationship to resource conditions and dynamics. In summarizing the potential contributions of social science to ecosystem management, Endter-Wada et al.[6] concluded that understanding spatial variability in resource needs, values, and uses was critical but highlighted a lack of systematic data analysis required to move beyond the rhetoric to the reality of integrating human values in ecosystem management. According to Grove et al.,[11] exploring questions about how motivations and capacities influence and are influenced by the biophysical environment will be best explored by adapting traditional social science field methods that have been applied to natural resource management.

Although numerous researchers have integrated nationally collected census data into landscape analyses, there are very few examples of attempts to purposefully collect social data that can be integrated with specific biophysical data layers. Brown[7] provides some insights into the application of these approaches as does earlier work by Curtis, Byron, and McDonald[12] and Curtis, Byron, and MacKay,[13] upon which this chapter aims to extend.

This chapter draws on findings from spatially referenced surveys of land managers to highlight methodologies for integrating social and biophysical data at a regional or catchment scale. Specific issues and approaches covered include mapping land use change and exploring the extent and nature of links between mapped biophysical resource conditions and land manager perceptions, values, and practices.

3.2 MAPPING CHANGE IN LAND USE AT A REGIONAL SCALE

3.2.1 Why Understanding Land Use Change Is Important

A capacity for detecting and reporting land use change is critical to evaluating and monitoring trends in natural resource conditions and the effectiveness of public investment in natural resource management. Australia's National Land and Water Resources Audit, for example, has identified the reporting of change over time and the integration of land use information with other natural resource information as a key to effectively addressing major sustainability problems such as salinity, water quality, and soil loss.

3.2.2 Types of Land Use Change

A particular difficulty with land use change reporting is discriminating the different dimensions of change. Protocols for reporting land use change in an agricultural context, for example, should be capable of distinguishing the temporal characteristics of farming systems (e.g., rotations), seasonal variability, and longer-term industry and regional trends. Lesslie, Barson, and Smith[14] identify four broad approaches to measuring and reporting land use change:

1. *Areal change: loss or gain in the areal extent.* This provides an indication of whether target land uses are increasing or decreasing in area over time. Changes can be presented statistically, graphically, or spatially and identified changes compared and trends observed.
2. *Transformation: the pattern of transition from one land use to another.* For example, an area may be cropped one year, grazed the next year, and then cropped again the year after. Alternatively, land under improved pasture for dairy may be permanently converted to vineyards.[15] Land use transformations between time periods may be expressed using a change matrix.
3. *Dynamics: rates of change and periodicity in areal extent or transformations.* The temporal nature of change may be further explored by analyzing whether rates of change are increasing or decreasing, are long- or short-term trends, or cyclic (for example, changes as a result of differences in growing seasons, structural adjustment, farming systems, or rotation regimes). This may reveal key trends in land use and land management not evident in expressions of simple areal change or transformations. Successful analysis of land use dynamics requires consistent, high-quality time-series data. Often it is not possible to obtain sufficiently consistent data over consecutive years or seasons.
4. *Prediction: modeling spatial or temporal patterns of change.* The use of models to predict past, present, and future land uses based on certain rules, relationships, and input data may help identify key drivers of land use change, implement scenario planning, and fill gaps in data availability.

3.2.3 Changes in Land Use in the Lower Murray Region

The capacity to report change also depends on the availability of consistent time-series data capable of providing insights into relevant aspects of change. Where fine-scaled time-series data are available, spatial analysis can provide important insight into the nature of land use change.

Using time-series data from fine-scaled mapping based on orthophoto interpretation and detailed property surveys, it is possible to highlight spatial trends in land use patterns. For example, Figure 3.1 shows trends in the expansion of irrigated horticulture around the towns of Renmark, Berri, and Loxton in the Lower Murray region of southeastern Australia produced by the Australian Collaborative Land Use Mapping Program (ACLUMP), a partnership of Australian and state government agencies producing coordinated land use mapping for Australia.[14] This time-series, 1990 to 2003, is drawn from 1:25,000 catchment-scale land use mapping completed

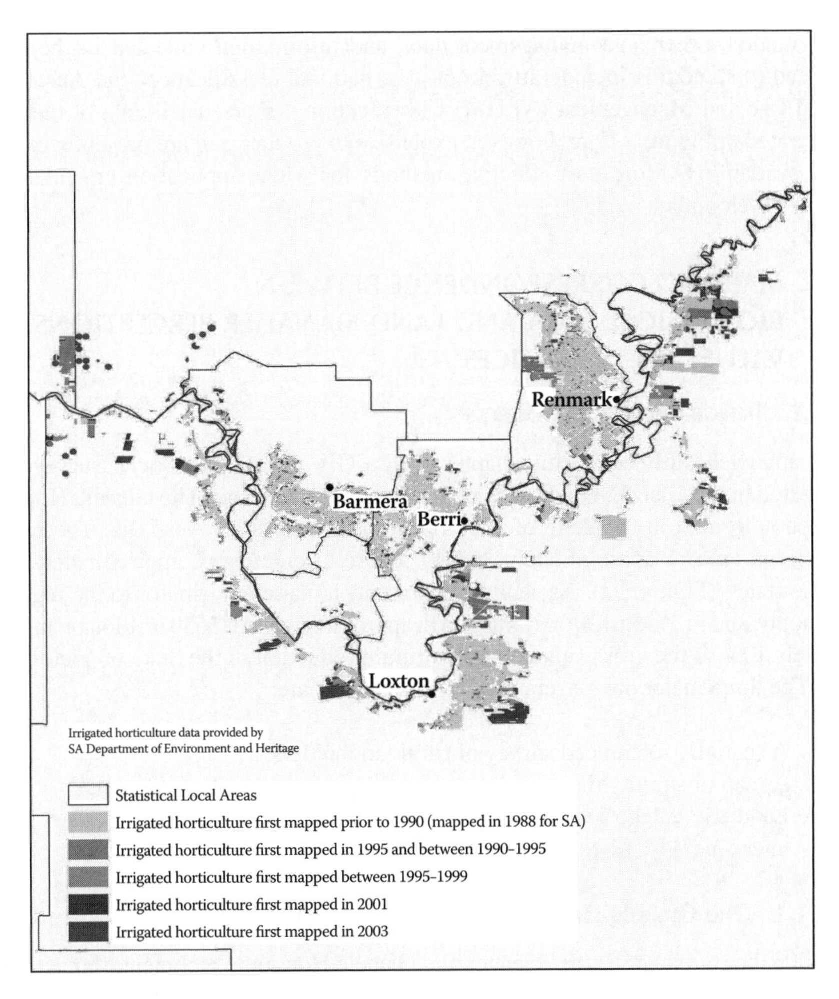

FIGURE 3.1 **(See color insert following p. 132.)** Land use change in the Barmera, Berri, and Renmark areas of South Australia.

using orthophoto interpretation and detailed property surveys.[16] The mapping reveals a pattern of land use transformation and intensification from dryland cereal cropping and grazing to irrigated horticulture, and a trend from small-scale enterprises clustering around town areas to dispersed, large-scale establishments at increasing distances from irrigation water supply (rivers).

Time-series land use mapping at catchment-scale in Australia is produced by ACLUMP to agreed to national standards, facilitating its use in national and regional natural resource assessments. The mapping process involves stages of data collation, interpretation, verification, independent validation, quality assurance, and the production of land use data and metadata. This includes collecting existing land use information and compiling it into a digital data set using a GIS. Important information sources include remotely sensed information, land parcel boundary information, forest and reserve estate mapping, land cover, local government zoning information, other land management data, and information collected in the field. Agreed-to standards include attribution to a national classification, the Australian Land Use and Management (ALUM) Classification.[14] Fine-scaled data of the type illustrated in Figure 3.1 are, however, expensive to produce and are presently of limited availability. More cost-effective methods for wider application are presently under development.

3.3 MAPPING CORRESPONDENCE BETWEEN BIOPHYSICAL DATA AND LAND MANAGER PERCEPTIONS, VALUES AND PRACTICES

3.3.1 Integrating Data Sources

The approach outlined in this chapter used a GIS to integrate social survey data collected in the Glenelg Hopkins region with biophysical data. The Glenelg Hopkins region is located in the State of Victoria in the southeast of Australia. The region covers an area of approximately 26,000 square kilometres or approximately 11% of the state[17] (Figure 3.2). Agriculture represents a major contributor to the regional economy, and in 1999 to 2000 it was worth approximately AU$650 million or approximately 10% of the gross value of agricultural production in the State of Victoria.

The three major data layers used in this chapter are:

1. A spatially referenced survey of rural landholders[18]
2. A map of salinity discharge based on the groundwater flow systems[19]
3. Land use categorized as Conservation of Natural Environment (Class 1) under the ALUM classification system[20]

3.3.1.1 The Glenelg Hopkins Landholder Survey

In 2003 the Bureau of Rural Sciences and Glenelg Hopkins Catchment Management Authority conducted a survey of approximately 1,900 rural landholders from across the region.[18] The survey focused on gathering baseline information regarding the key social and economic factors affecting landholder decision making about the adoption of practices expected to improve the management of natural resources in the Glenelg Hopkins region. The survey was sent to a random selection of rural property owners, with properties over 10 hectares in size, identified through local rate payer databases. A final response rate of 64% was achieved for this survey. All survey data (some 250 variables) were entered into a geographical information system (ArcView GIS) and assigned to a property centroid using *x* and *y* coordinates included in the rate payer databases.

3.3.1.2 Salinity Discharge Sites Based on Groundwater Flows Systems

The map of salinity discharge sites in the Glenelg Hopkins region was undertaken as part of the groundwater flow systems project conducted by Dahlhaus, Heislers, and Dyson.[19] The groundwater flow systems were developed by the National Land and Water Resources Audit as a framework for dryland salinity management in

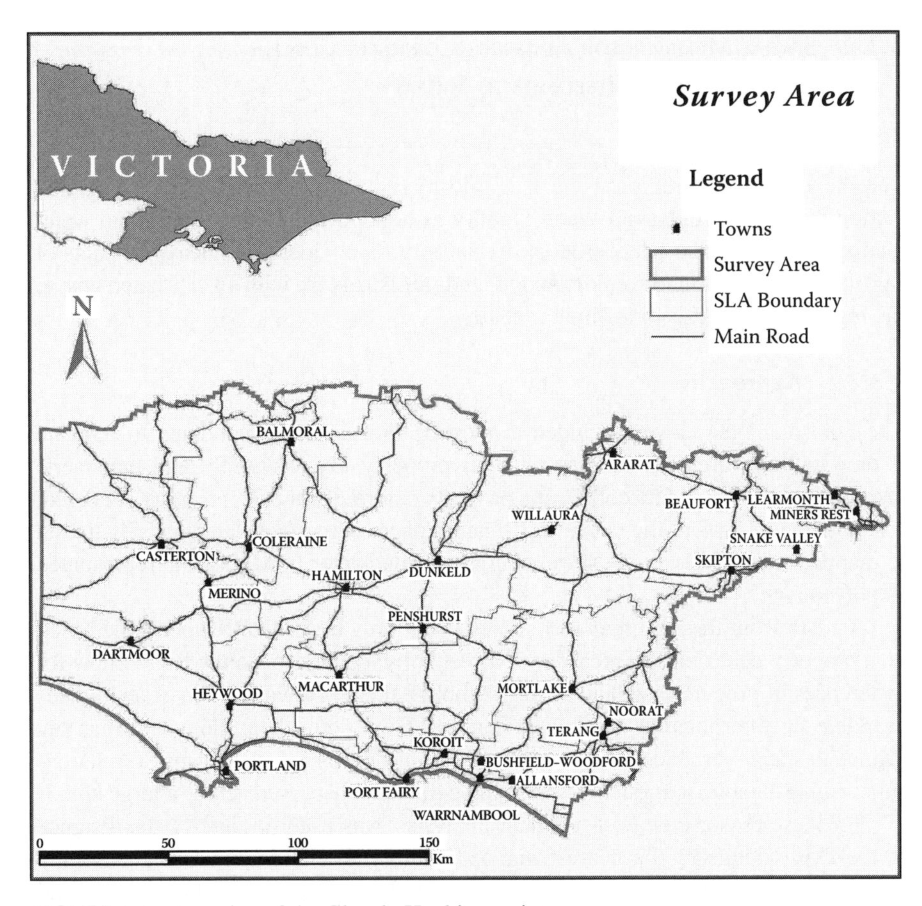

FIGURE 3.2 Location of the Glenelg Hopkins region.

Australia.[21] This work categorizes landscapes based on similarities in groundwater processes, salinity issues, and management options. Dahlhaus, Heislers, and Dyson[19] stated that while groundwater flow systems are a useful tool in helping to understand salinity, there has been little scientific validation of the flow systems or salinity processes in the Glenelg Hopkins region.

3.3.1.3 Land Use Categorized as Conservation of Natural Environment

Land use mapping for the Glenelg Hopkins region in the State of Victoria was undertaken using a three-stage process.[20] The first stage of mapping involved the collation of existing land use information, remotely sensed information (satellite imagery and aerial photography), and cadastre. Other important information sources were reserve estate data, land cover, local government zoning information, and other land management data. The second stage in the mapping process involved interpretation and assignment of land use classes according to the ALUM classification to create an initial draft land use map. The final stages of mapping included field verification, the editing of draft land use maps, and validation. The mapping is dated at 2001 and is produced at scales of 1:25,000 and 1:100,000.

3.3.2 Spatial Methods for Assessing Correspondence in Assessments and Responses to Salinity

3.3.2.1 Context

The Glenelg Hopkins region is one of 21 priority regions identified under the National Action Plan for Salinity and Water Quality as being affected by salinity and water quality problems. The Glenelg Hopkins Salinity Plan[22] identified heavy impacts of salinity on agriculture, the environment, and infrastructure with an estimated cost to the region of over AU$44 million annually.

3.3.2.2 Approach

The land manager survey included a question that asked respondents to indicate if they had any areas of salinity on their property. By assigning land managers' responses to the point data containing property centroids for each property surveyed, it is possible to explore the extent that land manager perceptions are spatially linked to mapped salinity discharge sites using the groundwater flow systems (represented as polygons).

As data from the land manager survey could only be joined to a point file based on a property centroid, any measure of direct correspondence would fail to allow for differences in property size and shape. Although there is a wide range of techniques available for interpolating continuous surfaces from point data, the extent that any change in social variables can be predicted by algorithms based on a spatial relationship is questionable, particularly where the points are dispersed across a large area.[23]

For these reasons, nearest neighbor analysis[24] was used to identify the distance to the closest edge of the nearest mapped salinity discharge site for each survey respondent. These distances can then be compared for respondents who said they had salinity on their property and those who did not or across a range of other variables.

Although interpolating surfaces from the land manager point data is problematic, creating a raster-based surface of distance from any grid cell to the nearest salinity discharge site provides a quick visual display that can be overlayed with the land manager perceptions and salinity discharge layers (Figure 3.3).

3.3.2.3 Analysis

The results of the nearest neighbor analysis clearly show that landholders in close proximity to mapped salinity discharge sites were significantly more likely to identify areas of salinity on their property (Table 3.1). With over half of all respondents within 0.5 km of a discharge site identifying salinity on their property, applying this methodology also suggests that most landholders are aware of salinity on their property.

By adopting the nearest neighbor technique it is also possible to explore the extent that land managers closer to mapped salinity discharge sites are more likely to be concerned about the impacts of salinity and undertaking practices expected to help mitigate salinity (Table 3.2).

Although most respondents close to mapped salinity discharge sites appear to be aware of the issue, there were still a large number of respondents near mapped

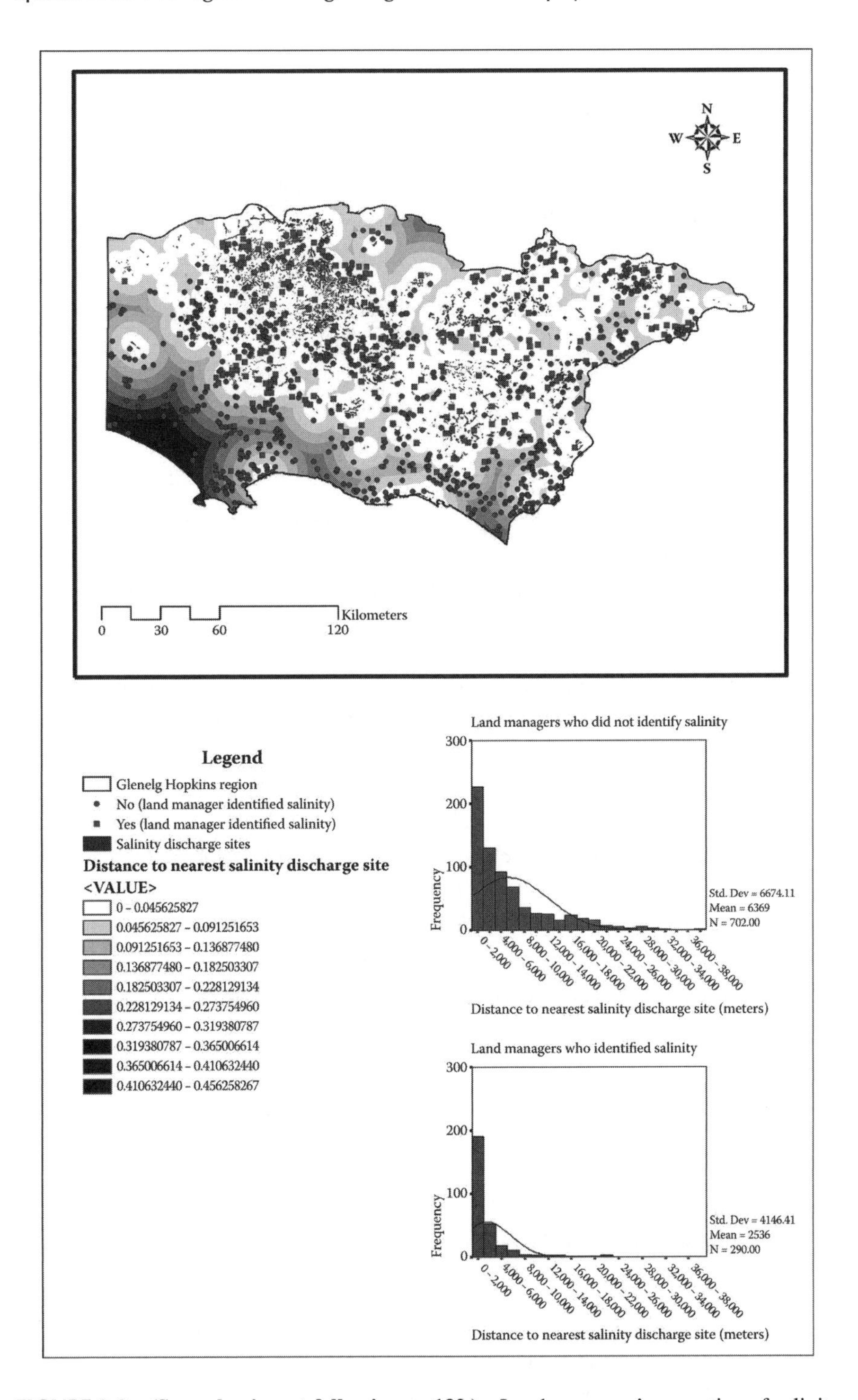

FIGURE 3.3 **(See color insert following p. 132.)** Land managers' perception of salinity and mapped salinity discharge sites.

TABLE 3.1
Land Managers' Perception of Salinity and Distance to Mapped Salinity Discharge

Distance to nearest mapped salinity discharge site (m)	Land manager identified salinity (%)	
	Yes	No
0–499	61	39
500–999	47	53
1,000–1,999	35	65
2,000–2,999	27	73
3,000–3,999	31	69
4,000–4,999	17	83
Over 5,000	11	89

TABLE 3.2
Distance to Mapped Salinity Discharge and Land Manager Attitudes and Practices

Land manager attitudes and practices	Median distance to nearest mapped salinity discharge site (m)
Concern about salinity reducing productive capacity of their property [a]	
High	1,639
Moderate	2,028
Low	3,092
Concern about salinity reducing productive capacity of the local area [a]	
High	1,951
Moderate	2,386
Low	3,202
Concern about salinity reducing water quality in the local area [a]	
High	1,824
Moderate	2,262
Low	3,216
Planted trees or shrubs [a]	
Yes	2,493
No	3,498
Established deep rooted perennial pasture [b]	
Yes	2,764
No	2,648

[a] Difference was statistically significant ($p < 0.05$) using the Kruskal-Wallis nonparametric chi-square test.

[b] Difference was not statistically significant ($p > 0.05$) using the Kruskal-Wallis nonparametric chi-square test.

TABLE 3.3
Differences between Land Managers Who Were Aware of Salinity and Those Unaware

Characteristics of land managers[a]	Land managers within 500 m of salinity	
	Aware	Unaware
Primary occupation farming	85%	56%
Member of a Landcare group	70%	37%
Completed a training course related to property management	69%	25%
Had work undertaken on their property that was at least partially funded by government	53%	19%
Planted trees and shrubs	85%	67%
Established perennial pasture	73%	44%
Median property size	465 ha	136 ha

[a] All differences were statistically significant ($p < 0.05$) using the Pearson chi-square test.

salinity that appear to be unaware of the problem. For the purposes of this example we have assumed that landholders within 0.5 kilometer of a discharge site that have not identified salinity on their property are unaware of the problem.

The spatial identification of respondents who appear to be unaware of salinity on their property provides an important opportunity to identify key characteristics of this group of respondents and thus develop better targeted community engagement strategies. For example, Table 3.3 clearly highlights a distinctive set of characteristics of landholders that appear to be unaware of salinity on their property.

3.3.3 Mapping the Relationships between Areas of High Conservation Value and Land Managers' Values and Practices

3.3.3.1 Context

A key aim for natural resource management in the Glenelg Hopkins region is to maintain and enhance remnant native vegetation. The Glenelg Hopkins region has an extensive history of land clearing, and native vegetation now covers less than 13% of the region, with 8% in parks and reserves fragmented across the region.

3.3.3.2 Approach

The combination of data collected through the regional landholder survey with land use mapping data for the Glenelg Hopkins region provides an opportunity to identify those land managers who are most likely to have an impact on areas of high conservation value.

Nearest neighbor analysis for each survey respondent (represented as a property centroid) to the nearest edge of an area classified as having high conservation value (polygon) can be used to help identify key groups of land managers. Once the distance to the nearest area of high conservation value has been computed, standard statistical analyses can be used to help discern patterns between spatial proximity and key characteristics of land managers and their property. As outlined previously, a raster-based surface of distance from any grid cell to the nearest area of high conservation value provides a useful overlay to help graphically represent relationships (Figure 3.4).

3.3.3.3 Analysis

When applied to survey and land use data from the Glenelg Hopkins region these analyses show some very clear differences in the characteristics of land managers and their property based on their proximity to areas of high conservation value (Table 3.4).

Applying the same technique it is also possible to explore if the values landholders attach to their property in terms of the social, economic, and environmental benefits are linked spatially to areas of high conservation value. These analyses show that respondents who said being close to nature was an important value of their property were in fact significantly closer to areas of high conservation value, while in turn those who said providing household income was important were significantly further from these areas. However, it is interesting to note that the property providing the sort of lifestyle desired was not linked to the proximity to areas of high conservation value (Table 3.4).

Finally, survey data and land use mapping data can be compared to see if land managers near areas of high conservation are more likely to have adopted management practices aimed at improving biodiversity. These analyses show mixed results. Land managers closer to areas of high conservation value were significantly more likely to have encouraged regrowth of native vegetation and fenced off areas of native vegetation on their property. However, these respondents were also significantly less likely to have planted native trees and shrubs on their property (Table 3.4).

3.4 INSIGHTS AND IMPLICATIONS FROM INTEGRATING SOCIAL AND BIOPHYSICAL DATA AT A REGIONAL SCALE

The use of spatial methodologies for integrating social and biophysical data, as demonstrated in this chapter, has some important insights and implications for efforts to improve natural resource management outcomes.

In the first instance, these approaches highlighted a number of important discrepancies between respondents' assessments of salinity and those made using the ground water flow system. Over a third of all respondents within 500 meters of a mapped salinity discharge site did not identify areas of salinity on their property, and over half of all the land managers who identified areas of salinity did not correspond with mapped salinity discharge sites. Many parts of the Glenelg Hopkins region are characterized by relatively high rainfall, and it is possible that some landholders have mistaken waterlogging as a sign of salinity. It is also possible that small localized areas of salinity identified by landholders could be missed through interpretation of large-scale aerial photographs. Similarly, the identification of salt indictor species through the interpretation of aerial photographs as part of the mapping of salinity

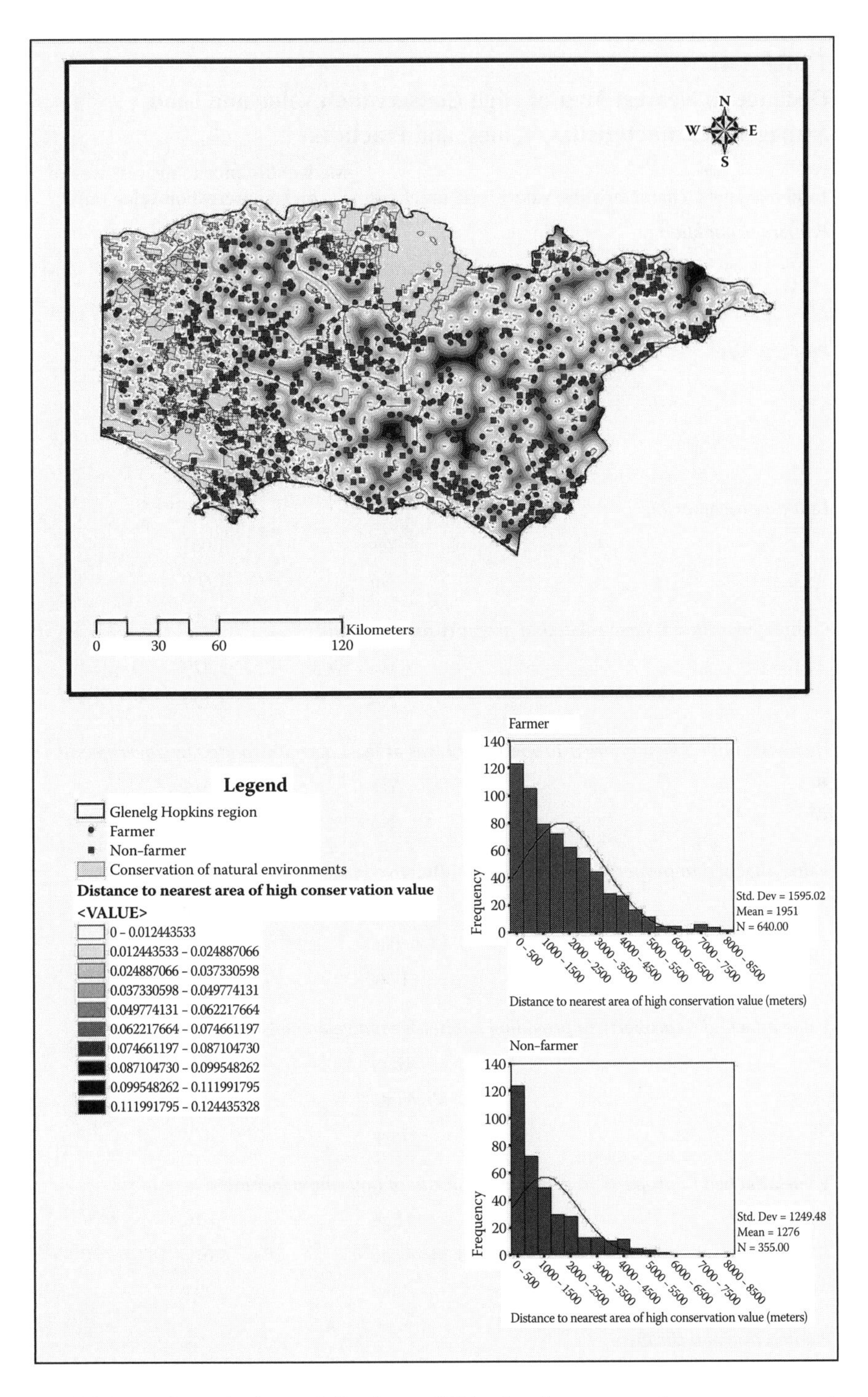

FIGURE 3.4 **(See color insert following p. 132.)** Land managers who manage properties near areas of high conservation value.

TABLE 3.4
Distance to Nearest Area of High Conservation Value and Land Managers' Characteristics, Values, and Practices

Land managers' characteristics, values, and practices	Median distance to nearest area of high conservation value (m)
Primary occupation[a]	
Farmer	1,614
Non-farmer	794
Property size[a]	
Small (< 100 ha)	829
Medium (100–499 ha)	1,425
Large (500 ha and over)	1,642
Landcare membership[a]	
Yes	1,063
No	1,600
Completed a short course related to property management[a]	
Yes	1,429
No	1,129
Had work undertaken on their property that was at least partially funded by government[a]	
Yes	1,512
No	1,152
Value attached to property in providing an attractive place to live[b]	
High	1,276
Moderate	1,470
Low	969
Value attached to property in providing habitat for native animals[a]	
High	1,074
Moderate	1,357
Low	1,479
Value attached to property in providing majority of household income[a]	
High	1,616
Moderate	1,012
Low	794
Planted tree and shrubs[a]	
Yes	1,470
No	881

TABLE 3.4 (continued)
Distance to Nearest Area of High Conservation Value and Land Managers' Characteristics, Values, and Practices

Land managers' characteristics, values, and practices		Median distance to nearest area of high conservation value (m)
Encouraged regrowth of native vegetation[a]		
	Yes	1,122
	No	1,352
Fenced areas of native bush[a]		
	Yes	856
	No	1,357

[a] Difference was statistically significant ($p < 0.05$) using the Kruskal–Wallis non-parametric chi-square test.

[b] Difference was not statistically significant ($p > 0.05$) using the Kruskal–Wallis non-parametric chi-square test.

discharge sites could also be confounded by high rainfall and a tendency for waterlogging. Incorporating data on rainfall and rainfall reliability across the region is one option that may help further identify reasons for conflicting perceptions of salinity.

To the extent that we assume that the salinity discharge maps have correctly identified areas of salinity, the integration of spatially referenced survey data provides important insights for the development of targeted management strategies. Through this process it is possible to spatially locate not only areas affected by salinity but areas within those, where the land managers are unaware of, or not actively managing, the problem. Taken in isolation either the maps of salinity discharge or landholder identified salinity are inadequate for targeted strategies to mitigate salinity. For example, if most landholders in a region are not affected or directly contributing to salinity, the capacity of land managers at an aggregate scale to modify land management practices is largely irrelevant. That is, rather than undertaking broad outreach activities about salinity, the methods presented in this chapter allow regional catchment managers to develop more strategic investments.

Similarly, linking remnant areas of vegetation through targeted investment is an integral part of plans to improve biodiversity outcomes in the Glenelg Hopkins region (Figure 3.5). The integration of land use mapping of areas of high conservation value with spatially referenced landholder information clearly highlighted that land managers near areas of high conservation areas were quite different and appear to preferentially purchase properties with high amenity value.

The fact that land managers who were unaware of salinity and those near areas of high conservation value tended to be nonfarmers, own smaller properties, were less likely to be involved in the Landcare program or have completed a course related to property management have important implications. The small numbers of large family farming operations that manage much of the land area of Australia have been a key target of policies and programs aimed at improving natural resource management. Although these programs and policies have been largely successful, as

FIGURE 3.5 (a) Pastures, (b) rolling country-side and (c) canola cropping in the Glenelg Hopkins region.

in the case of flagship programs such as Landcare, there is increasing evidence of a growing number of rural lifestyle motivated "tree-change" or subcommercial land holdings.[25] There is a clear need for regional catchment managers to better target and engage these land managers who are likely to operate outside their normal information channels and networks, particularly to the extent that they manage critical natural resources, as is the case in the Glenelg Hopkins region.

3.5 FUTURE DIRECTIONS

The examples presented in this chapter linking spatially referenced social survey data with biophysical data at a regional scale provides only a first glimpse at the potential of these approaches. Other applications may include exploring on-property profitability against rainfall reliability, cropping techniques against soil erosion and turbidity levels, or management of riparian zones and water quality.

More sophisticated techniques for mapping salt, such as airborne electromagnetics (AEM), also provide an opportunity to more precisely identify locations in a catchment where various management actions (such as planting trees) will help mitigate salinity. AEM provides a three-dimensional model of salt by measuring the electrical conductivity of the ground at different depths.[26] Furthermore, a recent study by Baker and Evans[27] using AEM found that tree planting in areas previously thought to be beneficial may actually contribute to salinity by reducing fresh water flows. The combination of AEM data with spatially referenced survey data appears likely to hold much promise in allowing better targeted and more site-specific approaches for managing dryland salinity.

Generally, the cost of land use change detection and reporting using fine-scaled time-series data based on orthophoto interpretation and detailed property surveys, as outlined in this chapter, is prohibitive and more cost-effective methods are needed. The capacity to adequately characterize change is highly dependent on matching the spatiotemporal accuracy and precision of available data to the relevant land use dynamic. The coupling of time-series satellite imagery to regularly collected agricultural statistics presents one practical and widely applicable approach to mapping agricultural land use change. A procedure adopted for regional-scale land use mapping in Australia by ACLUMP using agricultural census and survey data, Advanced Very High Resolution Radiometer (AVHRR) imagery, and a statistical spatial allocation procedure is promising for the development of annual time-series analyses, providing that limitations in spatial accuracy can be accommodated.[28,29] There is scope too for improvement if higher resolution satellite imagery (e.g., Moderate Resolution Imaging Spectroradiometer [MODIS]) can be applied successfully. ACLUMP partners are currently pursing investigations in this area.

3.6 CONCLUSIONS

This chapter has presented a range of simple and practical approaches for integrating spatially referenced social data with biophysical data layers. Furthermore, by presenting results from case studies demonstrating the application of these techniques, hopefully we have helped show how these approaches can contribute to improved management outcomes at a landscape scale.

The spatial methodologies outlined in this chapter are derived from our early attempts at creating meaningful integration between social and biophysical sciences, and barely scratch the surface of the many potential applications of this technique. In particular, the use of longitudinal survey data combined with longitudinal land use change data is likely to help more definitively understand the interactions between linked social and biophysical systems.

REFERENCES

1. Kenney, D. S. Are community-based watershed groups really effective?: Confronting the thorny issue of measuring success. *Chronicle of Community* 3(2), 33–38, 1999.
2. McGinnis, M. V., Woolley, J., and Gamman, J. Bioregional conflict resolution: Rebuilding community in watershed planning and organizing. *Environmental Management* 24(1), 1–12, 1999.
3. Curtis, A., Shindler, B., and Wright, A. Sustaining local watershed initiatives: Lessons from Landcare and watershed councils. *Journal of the American Water Resources Association* 38(5), 1–9, 2002.
4. Dale, A., and Bellamy, J. *Regional Resource Use Planning in Rangelands: An Australian Review*. Land and Water Resources Research and Development Corporation, Canberra, 1998.
5. Commonwealth of Australia. *NHT Annual Report 2003–2004*. Natural Heritage Trust, Canberra, 2004.
6. Endter-Wada, J. et al. framework for understanding social science contributions to ecosystem management. *Ecological Applications* 8(3), 891–904, 1998.
7. Brown, G. Mapping spatial attributes in survey research for natural resource management: Methods and applications. *Society and Natural Resource* 18, 17–39, 2004.
8. Commonwealth of Australia. *Australia's Natural Resources 1997–2002 and Beyond*. National Land and Water Audit, Canberra, 2002.
9. Radeloff, V. C. et al. Exploring the spatial relationship between census and land-cover data. *Society and Natural Resources* 13, 599–609, 2000.
10. Field, D. R. et al. Reaffirming social landscape analysis in landscape ecology: A conceptual framework. *Society and Natural Resources* 16, 349–361, 2003.
11. Grove, J. M. et al. Data and methods comparing social structure and vegetation structure of urban neighbourhoods in Baltimore, Maryland. *Society and Natural Resources* 19, 117–136, 2006.
12. Curtis, A., Byron, I., and McDonald, S. Integrating spatially referenced social and biophysical data to explore landholder responses to dryland salinity in Australia. *Journal of Environmental Management* 68, 397–407, 2003.
13. Curtis, A., Byron, I., and MacKay, J. Integrating socio-economic and biophysical data to underpin collaborative watershed management. *American Journal of Water Resources Association* 41(3), 549–563, 2005.
14. Lesslie, R. Barson, M., and Smith, J. Land use information for integrated natural resources management—a coordinated national mapping program for Australia. *Journal of Land Use Science* 1(1), 1–18, 2006.
15. Flavel, R., and Ratcliff, C. *Mount Lofty Ranges Implementation Project: Evaluating Regional Change*. Primary Industries and Resources South Australia, Adelaide, 2000.
16. Smith, J., and Lesslie, R. *Land Use Data Integration Case Study: The Lower Murray NAP Region*. Unpublished report to the National Land and Water Resources Audit. Bureau of Rural Sciences, Canberra, 2005.

17. Glenelg Hopkins CMA. *Glenelg Hopkins Regional Catchment Management Strategy.* GHCMA, Hamilton, 2003.
18. Byron, I., Curtis, A., and MacKay, J. *Providing Social Data to Underpin Catchment Planning in the Glenelg Hopkins Region.* Bureau of Rural Sciences, Canberra, 2004.
19. Dahlhaus P., Heislers, D., and Dyson P. *GHCMA Groundwater Flow Systems.* Dahlhaus Environmental Consulting Geology, GHCMA, Hamilton, 2002.
20. Bureau of Rural Sciences. *Guidelines for Land Use Mapping in Australia: Principles, Procedures and Definitions,* 3rd ed. Australian Government Department of Agriculture, Fisheries and Forestry, Canberra, 2006.
21. National Land and Water Resources Audit. *Australian Dryland Salinity Assessment 2000. Extent, Impacts, Processes, Monitoring and Management Options.* NLWRA, Canberra, 2001.
22. Glenelg Hopkins CMA. *Salinity Plan.* GHCMA, Hamilton, 2002.
23. Sinischalchi, J. M. et al. Mapping social change: A visualisation method used in the Monongahela national forest. *Society and Natural Resources.* 19, 71–78, 2006.
24. Jenness, J. Nearest features (nearfeat.avx) extension for ArcView 3.x, v. 3.8a. Jenness Enterprises. (Available at http://www.jennessent.com/arcview/nearest_features.htm). 2004.
25. Aslin, H. et al. *Peri-urban Landholders and Bio-security Issues—A Scoping Study.* Bureau of Rural Sciences, Canberra, 2004.
26. Spies, B., and Woodgate, P. *Salinity Mapping Methods in the Australian Context.* Land and Water Australia, Canberra, 2004.
27. Baker, P., and Evans W. R. *Mid Macquarie Community Salinity Prioritisation and Strategic Direction Project.* Bureau of Rural Sciences, Canberra, 2002.
28. Bureau of Rural Sciences. *Land Use Mapping for the Murray-Darling Basin: 1993, 1996, 1998, 2000 Maps.* Australian Government Department of Agriculture Fisheries and Forestry, Canberra, 2004.
29. Walker, P., and Mallawaarachchi, T. Disaggregating agricultural statistics using NOAA-AVHRRNDVI. *Remote Sensing of Environment* 63, 112–125, 1998.

4 An Integrated Socioeconomic Study of Deforestation in Western Uganda, 1990–2000

Ronnie Babigumira, Daniel Müller and Arild Angelsen

CONTENTS

4.1 INTRODUCTION

The past 20 years has been a period of intensive statistical investigation into the causes of tropical deforestation, with the work of Allen and Barnes[1] commonly referred to as the article that kicked-off this effort. Yet there is surprisingly limited convergence on the basic question: "what drives deforestation?" There are a number of reasons for this. First, the simple fact is that the answer to this question is context specific—it is not the same constellation of factors that can explain deforestation across the tropics. Second, one can expect some researcher bias, in the sense that the answers provided reflect the researchers' background: geographical focus, discipline, political view, and so forth. Third, the variables included have differed greatly—often determined by whatever data are easily available. These factors have lead to different and even contradictory deforestation stories being told. One way toward a consensus would be better and more integrated and holistic methodologies. This book makes the case for the need and role for spatially integrated models of coupled natural and human systems in the contexts of study and management of land use.

This chapter is an empirical application of an integrated approach using data from Western Uganda. Our objective is to analyze the role that the context within which land use agents operate plays in their land use decisions. To do this we integrate spatially explicit socioeconomic and biophysical data as well as data on land cover changes derived from remote sensing to estimate an econometric model of deforestation.

We argue like others that deforestation is mainly a result of actions of agents responding to incentives. Indeed, over the past 20 years most analysts have argued that tropical deforestation occurs primarily for economic reasons, that is, land users convert forest to nonforest uses if the new land rent they can get is higher than for forest uses. This approach is based on the fact that people and social organizations are the most substantial agents of change in forested ecosystems throughout the world.[2] Although this perspective is important, it is not the complete story of tropical deforestation. The incentives (land rent) are determined by the context within which agents operate, and a more comprehensive analysis needs to incorporate these as well.

Following a broad review of economic models of deforestation, Angelsen and Kaimowitz[3] recommended incorporation of agricultural census and survey data into a geographic information systems framework. They argued that models that combine remote observations with ground based social data would allow modelers to take into account the role of socioeconomic factors and have potential to improve our understanding of the determinants of land cover changes.[3,4]

This chapter introduces three key aspects of context, namely the socioeconomic, spatial, and institutional aspects. After a brief background on Uganda and the deforestation debate, we present a framework of analysis and then data and methods. The key results are then presented and discussed.

4.2 BACKGROUND

4.2.1 UGANDA

Uganda is a landlocked country covering about 236,000 km^2, 81% of which is suitable for agriculture owing to a rich endowment of soils and a climate that is generally favorable for farming throughout the year.[5,6,7] Uganda is to a large extent dependent on natural resources because the majority of Ugandans live in the rural areas with low-input low-output agriculture as the main source of livelihood.[7,8]

The country has enjoyed an impressive economic growth rate since the early 1990s, among the highest in Sub-Saharan Africa. This is in sharp contrast to its recent past. The late 1970s and the early 1980s were characterized by economic chaos that resulted from the civil unrest of the period. Macro- and microindicators of economic health were poor, with low savings rates, high inflation rates, and a high external debt burden. A tipping point in this trend, however, was the change in government in 1986. The new government then embarked on a number of initiatives to rehabilitate, stabilize, and expand the economy. The result of these initiatives was the onset of Uganda's own roaring nineties. The exception to this picture is the northern part of the country, where political instability and violence have emptied the countryside in many districts. It is for this reason that we do not focus on the whole country.

Additionally, population has been growing at an average of 2.5% per year,[9] almost doubling in just 22 years from 12.6 million in 1980 to about 24.7 million in 2002. During the latter part of this period growth was even higher, with an average growth rate of 3.4% between 1991 and 2002.[10] The population is projected to increase to 32.5 million by 2015.[7]

Given the high dependence on natural resources, the combination of economic and population growth will undoubtedly exert a lot of pressure on these resources. Uganda therefore provides an interesting study into how these socioeconomic dimensions could have impacted deforestation (Figure 4.1).

4.2.2 The Forestry Sector

Prior to the late 1990s, the extent of Uganda's forest estate was based on educated guesses. Lack of comprehensive data limited the determination of forest area and rates of deforestation. Initial estimates by the Food and Agriculture Organization (FAO) put the forest and woodland cover at 45% of the total land cover in 1890. More recent figures have been in the 20% to 25% range. Forest and woodland are important because only 3% of Ugandan households in rural areas and 8% in urban areas have access to grid electricity; the rest rely on biomass for energy sources.[11] It is estimated that forests provide an annual economic value of $360 million (6% of GDP). Trees through fuel wood and charcoal provide 90% of the energy demands with a projection of 75% in 2015.

The first effort to map Uganda's original vegetation was done by Langdale-Brown[12] in 1960 who estimated the extent of forest cover for 1900, 1926, and 1925.[13] These data show an increasing trend in the annual rate of change of forest cover (Table 4.1).

The next effort to map Uganda's forest estate was undertaken by Hamilton.[13] Using satellite imagery, Hamilton tried to map out clear standing forest. Our understanding of this map is that it focuses on what is subsequently referred to as tropical high forest by the National Biomass Study (NBS). The map reveals that forest is not a particularly common type of vegetation in Uganda. This led Hamilton to conclude that visions of vast sweeps of mahogany-rich jungles, such as are entertained by some planners, were quite illusory.

A more recent and comprehensive attempt was undertaken by NBS in a project started in 1989 with the objective of providing unique information on the distribution and indirectly consumption of woody biomass in the country.

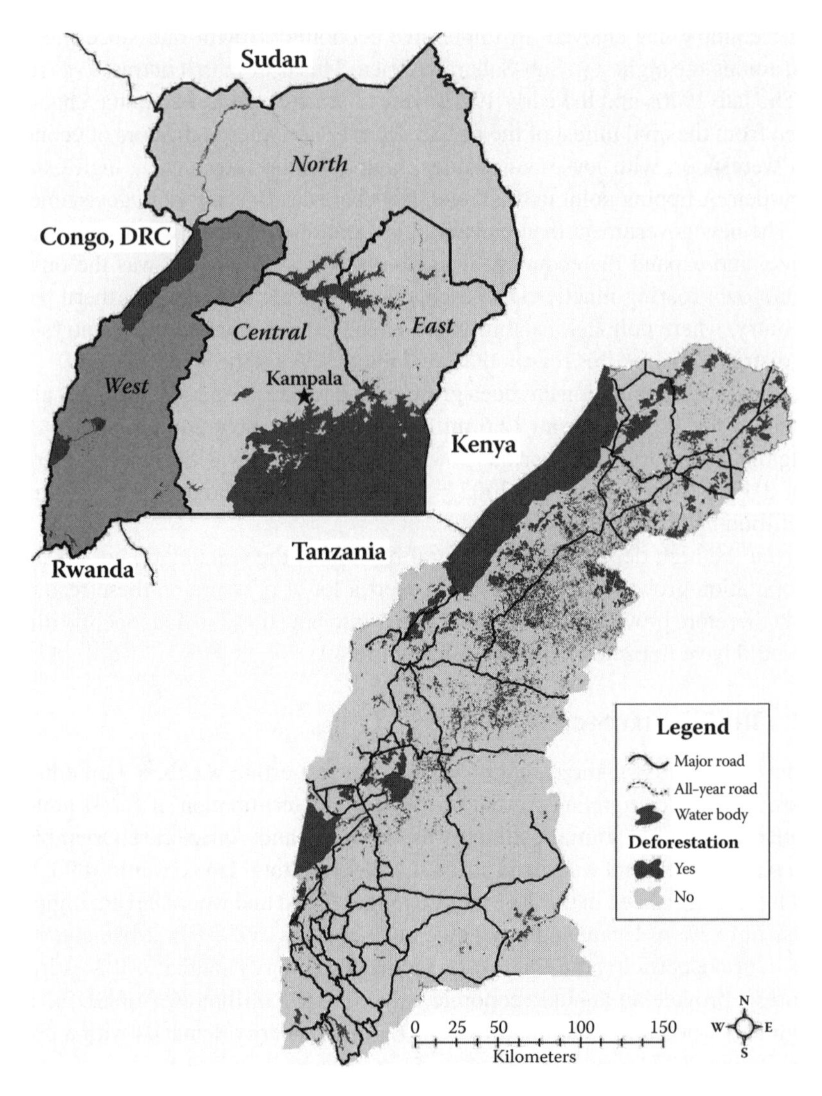

FIGURE 4.1 **(See color insert following p. 132.)** Uganda study area showing the distribution of deforestation within the western region of the country.

4.3 DEFORESTATION

4.3.1 Definitions of Deforestation

Deforestation has been used to describe changes in many different ecosystems. It is generally defined as loss of forest cover or forested land,[1,14] while Van Kooten and Bulte[15] define it as the removal of trees from a forested site and the conversion of land to another use, most often agriculture. FAO applies a similar definition—a permanent change from forest to nonforest land cover, with forest being defined as an area of minimum 0.5 ha with trees of minimum 5 m height in situ, minimum 10% canopy cover, and the main use not being agriculture.

TABLE 4.1
Early Estimates of Forest Cover and Deforestation Rates

Year	Forest and moist thicket (Ha)	Total area (%)	Annual forest loss[a] (HaY^{-1})
1900	3.1×10^6	12.7	
1926	2.6×10^6	10.8	1.8×10^4
1958	1.1×10^6	4.8	4.7×10^4

[a] Own calculations.
Source: Langdale-Brown (1960).[12]

More detailed definitions take into account what happens to the deforested land, transitions among classes, the magnitude of change, the threshold in area above which deforestation is said to have occurred, as well the temporal dimensions of the change.[16,17] As the precision in definition increases, so does the level of complexity and the challenges of empirical work. However, even recognizing the importance of exact definitions, the case for precision should not be exaggerated. Causes of major undesirable forest interventions can be analyzed and practical implications for policy making derived, even in a world with a relative lack of pure conceptual definitions.[18]

4.3.2 Good or Bad Deforestation

The debate on deforestation centers on whether tropical deforestation is an impending environmental disaster, one which if not addressed would have dire environmental consequences, or is just another overhyped agenda by environmentalists and some alarmist researchers.

For the *ever-worsening* school of thought, tropical deforestation is considered to be a major environmental crisis, because of its international impacts on biodiversity loss and climate and because of its local impacts such as an increase in flood occurrence, the depletion of forest resources, and soil erosion.[19] Such fears about the imminent extinction of large numbers of plants and animals have prompted an outpouring of concern and analysis about tropical deforestation in the past two decades.[20]

However, there is an *it's-not-that-bad* school that is a less pessimistic school arguing that there are no grounds for the alarmist claims.[21] Proponents of this school would go on to argue that deforestation is a natural, beneficial component of economic development especially in developing countries and is therefore nothing more than a gradual human alteration of an abundant natural resource (land) in order to increase productivity and welfare.

The former school is generally more prominent, owing to the visibility of the impacts of changes in local and international climate, and has resulted in the emergence of the social movement devoted to reducing deforestation. Important questions therefore remain about why, despite the emergence of this and the publication of hundreds of studies that analyzed its causes, the destruction of tropical rain forests did not appear to slow down much, if at all, during the 1990s.[20]

4.3.3 A Conceptual Framework

Deforestation is the result of two broad sets of processes: natural and human induced processes. In the former, forest reduction is induced by biotic and abiotic growth reducing factors within the forest ecosystem or as a result of broad climatic changes or catastrophes such as fires and land slides.[1] These natural processes, however, are often so slow and subtle as to be imperceptible.

On the other hand, the changes initiated by human activity tend to be rapid in progression, drastic in effects, widespread in scale, and thus more relevant to us on a day-to-day basis. Understanding the relationship between human behavior and forest change therefore poses a major challenge for development projects, policymakers, and environmental organizations that aim to improve forest management.[22]

To shed some light on this relationship, we take as our starting point, as have other models of deforestation in the von Thünen (1826) tradition, that any piece of land is put into the use that has the highest net benefits or land rent. The center of the discussion is then how various factors determine and influence the rent accrued from forest versus nonforest uses, and thereby the rate of deforestation. A recent extensive review of this approach is given by Angelsen.[23]

This approach is operationalized by modeling an agent (land use decision maker) living at or with access to the forest margin, whose aim is to maximize the land rent. (We are mindful of the pitfalls of applying a profit maximizing approach to rural households; however, we still believe this approach is informative.) Agents are individuals, groups of individuals, or institutions that directly convert forested lands to other uses or that intervene in forests without necessarily causing deforestation but substantially reduce their productive capacity. They include shifting cultivators, private and government logging companies, mining and oil and farming corporations, forest concessionaires, and ranchers.[18] The main culprit or agent is generally thought to be the agricultural household dwelling at the forest frontier (this setting is plausible in Uganda given the dependence on forests for energy highlighted above).

The agent's decisions are influenced by a number of factors such as prices of agricultural outputs and inputs, available technologies, wage rates, credit access and conditions, household endowments, forest access (both physical and property rights), and biophysical variables like rainfall, slope, and soil suitability. Location, the center of attention in von Thünen's original work, does influence a number of these variables (e.g., prices and wage rates). These factors affect the agent's decisions directly and are, therefore, referred to as decision parameters or immediate causes of deforestation (cf. the terminology used by Angelsen and Kaimowitz[3]).

At the next level is the context within which the agents operate. These contextual forces determine deforestation via their impact on the decision parameters. These causes are more fundamental and often distanced in the sense that it is difficult to establish clear links between this set of factors and deforestation. They are a complex dynamic mix of the socioeconomic, spatial, and institutional systems of communities representing the fundamental organization of societies and interacting in ways that are difficult to predict. The above discussion can be summarized in Figure 4.2.

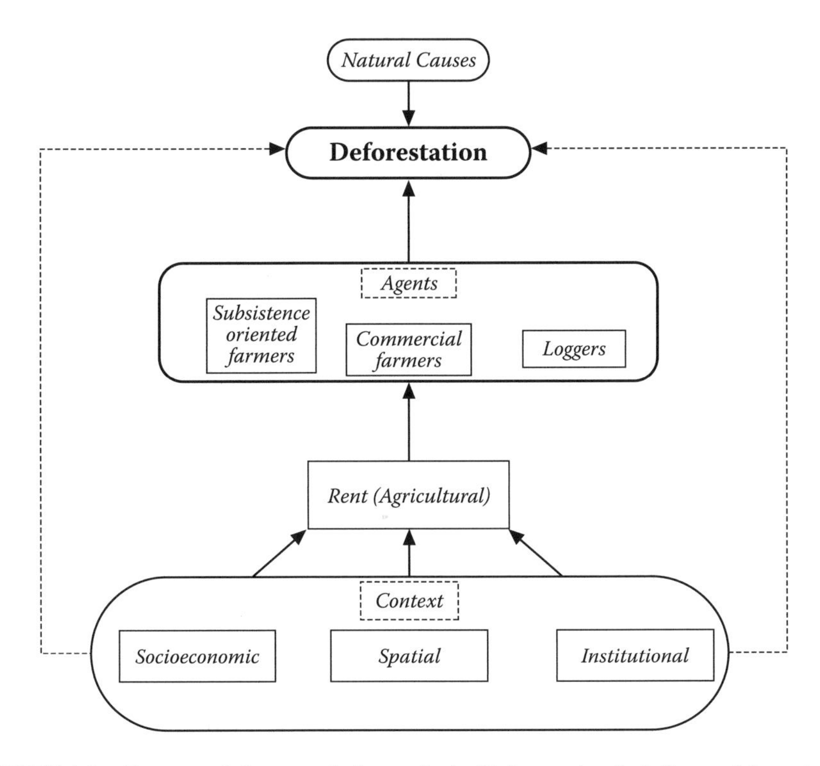

FIGURE 4.2 Conceptual framework for analysis. Deforestation is influenced by natural causes and human activities. The human activities are driven by the rental cost of land within socioeconomic, spatial, and institutional contexts.

4.4 DATA AND METHODS

4.4.1 DATA SOURCES

Land use and land cover data for this study come from land use/cover maps from the Uganda NBS and FAO Africover. Although we refer to them as the 1990 and 2000 maps, the satellite images used in their production are from 1989 to 1992 and 2000 to 2001, respectively, owing to the need to use cloud-free images.

The 1990 map was produced by visual interpretation of Spot XS satellite imagery from February 1989 to December 1992. Following preliminary interpretation, the map was verified through systematic and extensive ground truthing. The 2000 map is the FAO Africover land cover map produced from visual interpretation of digitally enhanced Landsat Thematic Mapper (TM) images (Bands 4, 3, 2) acquired mainly in the year 2000/2001. The land cover classes were developed using the Food and Agriculture Organization/United Nations Environmental Program (FAO/UNEP) international standard (LCCS) land cover classification system. The 2000 map was reclassified by staff at NBS to enable comparison between the two maps.

Administrative boundaries, infrastructure, and river maps come from the Department of Surveys and Mapping, Ministry of Lands, Housing and Urban Settlements and the Department of Surveys and Mapping. Socioeconomic data are from the National Population and Housing Census 1991, by the Statistics Department, Ministry of Finance and Economic Planning.

The slope and elevation were calculated from the digital elevation data of the Shuttle Radar Topographic Mission (SRTM) (CGIAR-CSI SRTM). Void-filled seamless SRTM data V1, accessed January 2005, available from the CGIAR-CSI SRTM 90m Database: http://srtm.csi.cgiar.org. Soil data are from Uganda's agroecological zones (AEZ) database[24] and from the results of a soil reconnaissance survey.[25] Following consultations with one of the authors of this map, we use soil organic matter and soil texture as the variables to capture soil suitability. We then calculate a weighted index from both raster maps. This index acts as a proxy for agricultural potential inherent in a parcel.

The different maps were projected into Universal Transverse Mercator (UTM) Zone 36, south of the equator and then assembled in a raster geographical information system (GIS) where we resampled the data to a common spatial resolution of 250 m. The choice of resolution was primarily guided by the need for a manageable data size.

A GIS was used to generate additional spatial variables, specifically the cost-adjusted distance to roads, the euclidean distance to water, and the euclidean distance to protected areas. We then export all the grids as ASCII files and import them into Stata 9,[26] which we use to carry out the descriptive and econometric analysis.

4.4.2 Econometric Model

To analyze the role that context plays in land use change, we estimate an econometric model for the probability deforestation. Our unit of analysis is a 6.25 ha pixel. Underlying this econometric model is a latent threshold model based on the idea that the land use decision regarding the parcel is made by an operator who can be a single person, household, or group of people in the case of common property ownership.[27] This operator may or may not own the parcel (our data does not allow us to make that distinction). However, we assume that for any given parcel, there is an operator who is able to make a land use decision pertaining to this parcel. A parcel will be cleared if it is economically profitable. That is:

$$R_{nft+1|f} \geq R_{ft+1|f}$$

where $R_{nft+1|f}$ represents the present value of the infinite stream of net returns from converting a parcel that was originally under forest (f) in period t to nonforest (nf) land use in period $t + 1$, which we will refer to as agricultural rent. This type of model is further discussed elsewhere.[27,28] In line with this integrated approach, the economic profitability of a parcel is a function of three sets of factors: the socioeconomic, spatial, and institutional contexts.

1. The *socioeconomic context* within which the parcel is embedded has a bearing on output prices and input costs. Higher output prices will increase agricultural rent, while higher wages translate into higher input costs, which reduce the rent and may thus reduce the probability of deforestation.

We argue that because the opportunity cost of labor in poor communities is typically very low, the probability of deforestation will be higher in poorer communities. Moreover, inequality may have a bearing within this framework. For any given average income, higher inequality implies a larger proportion of the population has an opportunity cost of labor below the level that makes forest clearing profitable. Thus we hypothesize that high inequality will be correlated with higher probabilities of deforestation.

2. The *spatial context* has an influence on the agricultural land rent. Included in this is the in situ resource quality, that is, the response of the land to the use without regard to its location determines the quantity of agricultural harvest possible from a given parcel, which in turn affects the probability of clearance. Also included is the accessibility and, by extension, all costs and benefits associated with a specific location as opposed to resource quality as well as idiosyncratic location-specific characteristics of the parcel. More accessible parcels are more likely to be cleared, and this does not necessarily mean that agriculture will be the subsequent land use. These parcels will be cleared mainly for the sale of timber.
3. Finally, the *institutional context* within which the agents operate also has an influence on agricultural land rent. This primarily refers to the property rights regimes in the communities that determine access and use rights. To the extent that they are enforceable, restrictions on clearance translate into a cost and thereby lower agricultural rent.

We therefore select a number of explanatory variables that best capture the context surrounding the management of the parcel. The variables and their origins are described in Table 4.2 together with our a priori expectations on their relationship with the likelihood of deforestation.

Our focus is on agricultural rent only, while forest rent is ignored. This simplification can be justified on two grounds: First, much of the forest is of de facto open access and the forest rent therefore is not captured by the individual land user (unlike agricultural rent). Second, during early stages in the forest transition (characterized by high levels of deforestation, such as in Western Uganda), changes in agricultural rent rather than forest rent are the key driver (cf. Angelsen[23]).

4.4.3 Methodological Issues

Conventional statistical analysis frequently imposes a number of conditions or assumptions on the data it uses. Foremost among these is the requirement that samples be random. Spatial data almost always violate this fundamental requirement, and the technical term describing this problem is *spatial autocorrelation.*[29]

Spatial autocorrelation (dependence) occurs when values or observations in space are functionally related. Spatial autocorrelation may arise from a number of sources such as measurement errors in spatial data that are propagated in the error terms or from interaction between spatial units. It may also arise from contiguity, clustering, spillovers, externalities, or interdependencies across space.

TABLE 4.2
Description of Variables

Variable	Description	Source	Expected sign[a]
	Socioeconomic Context		
head_emp	Employed household heads	Census 91	–
educ_Gini	Education Gini coefficient[b]	Census 91	+
popdens	Population density	Census 91	+
mig_share	Share of migrants in parish	Census 91	+
	Spatial Context		
cdcity_allrds	Cost adjusted distance to roads	Infrastructure map	–
dwater	Distance to water	Infrastructure map	–
slp	Slope	DEM	–
elev	Elevation	DEM	–
soil+2cl	Proportion of suitable soils	CIAT	+
rain	Rainfall	CIAT	?
x	Latitude index value	LUC & infrastructure maps	?
y	Longitude index value	LUC & infrastructure maps	?
	Institutional Context		
dprotect	Distance from protected areas	LUC & infrastructure maps	?
prtct	Protected area dummy	LUC & infrastructure maps	–

[a] *A priori* expectations on the effect of variables on deforestation (–) less; (+) more deforestation; (?) ambiguous. CIAT, International Center for Tropical Agriculture; DEM, digital elevation model; LUC, land use cover.

[b] The Education Gini coefficient is a measure of inequality ranging from zero (perfect equality) to one (perfect inequality).

Three approaches for correcting for spatial effects are often mentioned in the literature: regular sampling from a grid, pure spatial lag variables using latitude and longitude index values, and spatial lag variables involving a geophysical variable such as a slope or rainfall.[30]

Before carrying out the econometric estimation, we test for spatial dependence using the *SPDEP* package[31] in *R* language.[32] We find evidence of spatial autocorrelation at both the pixel and parish levels. We minimize the effects of spatial autocorrelation by including latitude and longitude index variables, and by drawing a sample from a grid with a distance of 500 m between cells.

4.5 RESULTS AND DISCUSSION

4.5.1 Descriptive Statistics

Most deforestation was concentrated in a few areas. A plot of cumulative distribution of deforestation shows that 15% of the parishes accounted for 70% of the total deforestation (Figure 4.3). Furthermore, most of the deforestation (60%) was within 10 km from main roads (Figure 4.4).

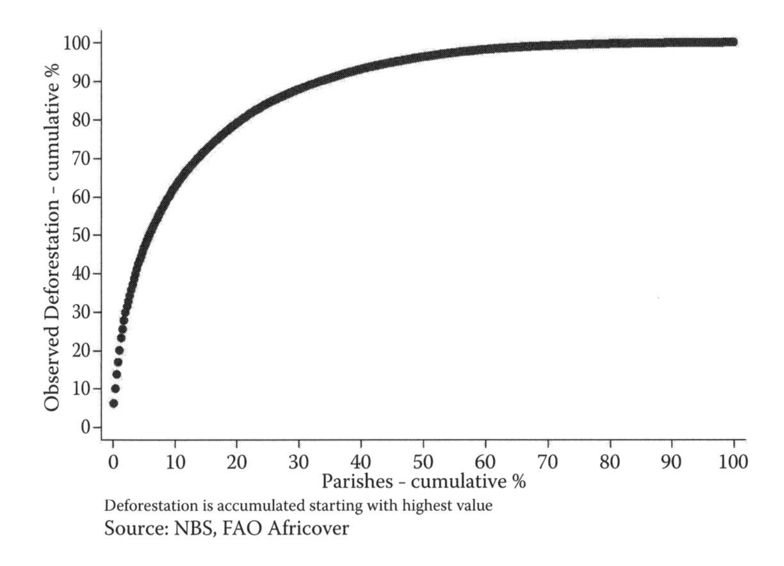

FIGURE 4.3 Cumulative deforestation by parishes across Uganda starting with the largest percentage area.

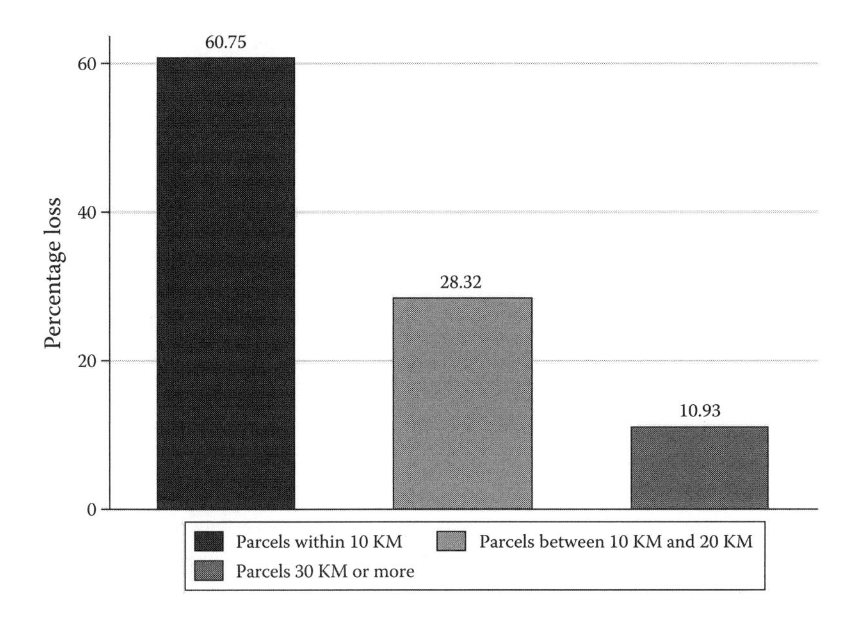

FIGURE 4.4 Relationship between deforestation and distance from roads.

Some descriptive statistics are presented in Table 4.3. Compared to all other parcels, forest parcels had more rainfall, were at a lower elevation, and were on less steep slopes. Not surprisingly, forest parcels were generally located farther from urban centers and farther from the main roads. This is typical of a von Thünen development process, with areas close to urban centers being cleared first and

TABLE 4.3
Descriptive Statistics

	All Parcels		
		(N = 697,060)	
	Mean	**Minimum**	**Maximum**
Proportion of suitable soils	64.89	32	91
Rainfall (mm)	1,091.05	701	1,949
Elevation (meters)	1,296.52	601	4,391
Slope	6.01	0.00	63.66
Cost-adjusted distance to roads	0.51	0.00	1.67
Distance to urban centers (km)	64.68	0.00	167.38
Distance to road (km)	8.74	0.00	38.57
Distance from protected areas (km)	6.83	0.00	38.21
Distance from water (km)	16.29	0.25	57.34
	Forest Parcels		
	All	**Deforested**	**Nondeforested**
	(N = 194,601)	**(N = 46,420)**	**(N = 148,181)**
	Mean	**Mean**	**Mean**
Proportion of suitable soils	62.51	63.78	62.11
Rainfall (mm)	1,177.46	1,148.02	1,186.69
Elevation (m)	1,230.12	1,052.06	1,285.91
Slope	5.2	3.77	5.65
Cost-adjusted distance to roads (km)	0.65	0.62	0.66
Distance to urban centers (km)	82.91	85.97	81.95
Distance to main road (km)	9.98	8.97	10.29
Distance from protected areas (km)	3.23	4.78	2.75
Distance from water (km)	19.3	18.28	19.62

expanding as population and the economy grows. Most forest parcels were in or close to protected areas.

Within the forested parcels, the ones that were deforested were at a lower elevation with less steep slopes. This suggests that accessibility was a key factor in the decision to clear a forest parcel, consistent with our explanation above.

4.5.2 Econometric Results and Discussion

Given the binary nature of the dependent variable, that is, the land is either cleared or it is not, we estimate a binary logit model. We correct for possible correlation in the error terms of pixels within a parish and use the Huber and White sandwich estimator to obtain robust variance estimates.

The econometric results are presented in Table 4.4. The dependent variable is a categorical variable indicating whether or not the parcel was deforested. The

TABLE 4.4
Logit Model Results of Deforestation in Western Uganda, 1990–2000

Variable	Coef.	*p*-value	e^(b*sdx)[a]
	Socioeconomic Context		
Proportion of heads that are employed	0.828*	0.034	1.202
Education Gini coefficient	0.329	0.675	1.028
Population density (pp/ha)	0.291**	0.008	1.166
Share of migrants in parish	1.363**	0.008	1.302
	Spatial Context		
Cost adjusted distance to roads	–1.318**	0.004	0.709
Distance to water	–0.004	0.503	0.948
Slope	0.051***	0.000	1.366
Elevation	–0.002***	0.000	0.339
Proportion of suitable soils	–0.004	0.579	0.948
Rainfall	–0.002	0.054	0.784
X	–0.003***	0.000	0.450
Y	–0.003***	0.000	0.409
	Institutional Context		
Protected dummy	–1.912***	0.000	0.388
Distance from protected area	–0.003	0.843	0.985
Constant	8.250***	0.000	
No. of parcels	43,760		
Model *p*-value	0.000		
Pseudo R^2	0.191		

[a] Change in odds for one SD increase in x.
Dependent variable = 1 if parcel changed from forest to nonforest class, 0 otherwise.
Percentage of correct predictions = 76.3.
*, $p < .05$; **, $p < .01$; ***, $p < .001$.

regression model was significant at the $p = .001$ level. Most coefficients have the expected signs. Below we discuss the statistically significant results.

4.5.2.1 Socioeconomic Context

The results show that deforestation is more likely to occur in better-off communities. Our proxy for wealth in the community—proportion of household heads that are employed outside the farm—is statistically significant at the 5% level, with an odds ratio (for one standard deviation increase) of 1.2. There are two contradictory effects of the better-off farm employment opportunities reflected by this variable. First, good employment opportunities increase the opportunity costs of labor and thereby lower agricultural rent and reduce the pressure on forest conversion. But, a higher share of off-farm employment is also correlated with economic development, creating higher demand for agricultural products. Our results suggest that in the Western Ugandan context the latter effect is dominating.

Consistent with this explanation, we also find that higher population densities have a positive impact on forest conversion. A high share of migrants in the parish also pulls in the same direction. Migrants have initially no or very small parcels of agricultural land and can therefore be expected to become major agents of deforestation. Thus the empirical model suggests that migration to take advantage of forest land suitable for agricultural conversion plays a major role. Inequality, as measured by the educational Gini, does not appear to have any significant effect on the likelihood of deforestation.

4.5.2.2 Spatial Context

Parcels closer to roads were more likely to be deforested. The descriptive statistics have already shown this, and the econometric results confirm the importance of distance as a factor. This result suggests, in line with numerous other studies, that interventions that reduce the cost of access to forested land will increase the likelihood of deforestation.

Not surprisingly we find that parcels at lower elevation were more likely to be deforested, suggesting again that the spatial context had a bearing on the cost of access. A one standard deviation increase in the elevation reduced the odds of deforestation by 0.34. A surprising result is, however, the effect of slope on the probability of deforestation. We anticipated that the probability of deforestation would be negatively correlated with the slope (mainly owing to the higher costs associated with working at higher elevation and steeper slopes). However, we find instead that steeper slopes were more likely to be deforested, and we do not have a satisfactory explanation of this result. A plausible explanation could be that lower lands have been converted and the pressure may have shifted to the marginal lands. Support for this can be found in the argument that fragile lands in sub-Saharan Africa are facing a worsening social and environmental crisis.[33]

Distance to water, rainfall, and proportion of suitable soils were not statistically significant. One reason for the insignificance of the soil variable might be that the map on which this variable is based is rather coarse, and therefore does not capture the relevant local specific variation that may exist. The same may be true for rainfall.

4.5.2.3 Institutional Context

An interesting result that emerges from this work is the fact that institutional interventions seem to have mitigated deforestation. Parcels in protected areas were less likely to be deforested. This is also what one can observe on the maps and on the ground: the protected areas are indeed “greener.” The conservation areas have been backed by relatively strong enforcement at the local levels with punishment of violators.

In addition, we tested if the conservation led to more land being deforested outside the conservation areas, a kind of negative spill over effect. Our hypothesis would then be that, after controlling for other factors, land close to the protected areas should experience higher forest conversion. But this variable is not statistically significant thus we could not reject the null hypothesis.

4.6 CONCLUDING REMARKS

The process of land use change is driven by a complex web of factors that cuts across disciplines. This means that efforts to address the land use change process should similarly be holistic and cut across disciplines. This chapter is an example of how such a study could be empirically carried out. We argued that additional insights could be gained from integrating spatially explicit socioeconomic, institutional, biophysical, and land cover data. Increasing availability of high resolution spatial data means that such an approach is possible in most places. The fact that the variables used to capture the socioeconomic context are significant shows that such a framework can be of policy relevance, for example, by including the effect on natural resources in the design and implementation of human resettlement programs or infrastructure development projects.

Four main stories emerge from our econometric analysis: the poverty cum affluence, the population, the protection, and the spatial story. Although we set out thinking deforestation was driven by poor households, we do not find any evidence in support of this. Rather it appears that deforestation is more likely in the better-off communities. Second, high population densities, together with a high proportion of migrants who may be in greater need for agricultural land, has also played an important role for deforestation in Western Uganda during the 1990s (Figure 4.5).

FIGURE 4.5 Land use change in Uganda. (a) Forest clearing. (b) Banana plantations on cleared land. (c) Pastoral land use on cleared land.

Third, there is also a strong spatial story to this in terms of factors such as closeness to roads and low elevation, leading to more deforestation. Finally, our study shows that protection had been effective in reducing the likelihood of deforestation.

REFERENCES

1. Allen, J. C., and Barnes, D. F. The causes of deforestation in developing countries. *Annals of the Association of American Geographers* 75(2), 163–184, 1985.
2. Wear, D. N., Abt, R., and Mangold, R. People, space and time: Factors that will govern forest sustainability. In: *Transactions of the 63rd North American Wildlife and Natural Resources conference*. Wildlife Management Institute, Washington, D.C., 1998.
3. Angelsen, A., and Kaimowitz, D. Rethinking the causes of deforestation: lessons from economic models. *World Bank Research Observer* 14(1), 73–98, 1999.
4. Rindfuss, R. R., and Stern, P. C. Linking remote sensing and social science: The need and challenges. In: Liverman, D., et al., eds., *People and Pixels: Linking Remote Sensing and Social Science*. National Academy Press, Washington, D.C., 1–27, 1998.
5. Mukiibi, J. K. Agriculture in Uganda. In: Mukiibi, J. K., ed., *Agriculture in Uganda*. Fountain Publishers, Kampala, 2001.
6. Esegu, J. F. O. Forest tree genetic resources in Uganda. In: Mukiibi, J. K., ed., *Agriculture in Uganda*. Fountain Publishers, Kampala, 2001.
7. NEMA. *State of the Environment Report for Uganda 2000/2001*. Ministry of Water, Lands and the Environment, Kampala, 2000.
8. MAAIF and MFPED. *Plan for Modernisation of Agriculture: Eradicating Poverty in Uganda*. Kampala, 2000.
9. NEMA. *State of the Environment Report for Uganda 1998*. Ministry of Water, Lands and the Environment, Kampala, 1998.
10. MFPED. *Background to the Budget for Financial Year 2004/05*, Kampala, 2004.
11. MFPED. *Poverty Eradication Action Plan (2004/5–2007/8)*, Kampala, 2004.
12. Langdale-Brown, I., The Vegetation of Uganda. Uganda Department of Agriculture 2(6), 1960.
13. Hamilton, A. C. *Deforestation in Uganda*. Oxford University Press, Nairobi, 1984.
14. Geist, H. J., and Lambin, E. F. *What Drives Tropical Deforestation?* Louvain-la-Neuve, 2001.
15. Van Kooten, G. C., and Bulte, E. *The Economics of Nature: Managing Biological Assets*. Malden, MA, Blackwell, 2000.
16. FAO. *State of the Worlds Forests, 1997*. Food and Agriculture Organization of the United Nations (FAO), Rome, 1997.
17. Myers, N. *Conversion of Tropical Moist Forests: A Report Prepared by Norman Myers for the Committee on Research Priorities in Tropical Biology of the National Research Council*. National Academy of Sciences, Washington, D.C., 1980.
18. Contreras-Hermosilla, A. *The Underlying Causes of Forest Decline*. Center for International Forestry Research (CIFOR), Bogor, Indonesia, 2000.
19. Caviglia, J. L., and Kahn, J. R. Diffusion of sustainable agriculture in the Brazilian tropical rain forest: A discrete choice analysis. In: Orlove, B. S., ed., *Current Anthropology*. University of Chicago Press, Chicago, 2001.
20. Rudel, T. K. *Tropical Forests: Regional Paths of Destruction and Regeneration in the Late Twentieth Century*. Chichester, Columbia University, New York, 2005.
21. Lomborg, B. *The Skeptical Environmentalist: Measuring the Real State of the World*. Cambridge University Press, Cambridge, 2001.

22. Southworth, J., and Tucker, C. The influence of accessibility, local institutions, and socioeconomic factors on forest cover change in the mountains of western Honduras. *Mountain Research and Development* 21(3), 276–283, 2001.
23. Angelsen, A. Forest cover change in space and time: combining von Thünen and the forest transition. In: *World Bank Policy Research Working Paper*, World Bank, Washington, D.C., 2006.
24. CIAT. *Bean Database for Africa and Uganda's Agroecological zones (AEZ) Database.* CD-ROM, Ver. 1.0, Kampala, 1999.
25. Memoirs of the Research Division, Kawand Research Station. Department of Agriculture, Uganda, 1960.
26. StataCorp. *Stata Statistical Software: Release 9*, College Station, TX, 2005.
27. Nelson, G. C., and Geoghegan, J. Deforestation and land use change: Sparse data environments. *Agricultural Economics* 27(3), 201–216, 2002.
28. Munroe, D., Southworth, J., and Tucker, C. M. The dynamics of land-cover change in western Honduras: Spatial autocorrelation and temporal variation. Prepared for the 2001 AAEA Annual Meetings, 2001.
29. O'Sullivan, D., and Unwin, D. *Geographic Information Analysis.* John Wiley & Sons, Hoboken, NJ, 2003.
30. de Pinto, A., and Nelson, G. C. Correcting for spatial effects in limited dependent variable regression: Assessing the value of "ad-hoc" techniques. Presented at American Agricultural Economics Association Annual Meeting, Long Beach, CA, 2002.
31. Bivand, R. Implementing spatial data analysis software tools in R. *Geographical Analysis* 38(1), 23–40, 2006.
32. R Development Core Team. *R: A language and environment for statistical computing.* R Foundation for Statistical Computing, Vienna, Austria, 2005.
33. Babu, S., and Hazell, P. Growth, poverty alleviation, and the environment in the fragile lands of sub-Saharan Africa. In: *Strategies for Poverty Alleviation and Sustainable Resource Management in the Fragile Lands of Sub-Saharan Africa.* Deutsche Stiftung für Internationale Entwicklung, Kampala, 1998.

5 Modeling Unplanned Land Cover Change across Scales

A Colombian Case Study

Andres Etter and Clive McAlpine

CONTENTS

5.1 INTRODUCTION

Land use is the interaction between humans and the biophysical environment with cumulative impacts on the structure, function, and dynamics of ecosystems at the local, regional, and global levels of ecological organization.[1] Human impacts on the Earth's environment are leaving an increasing ecological footprint, which now threatens many global ecosystems.[2,3] Human activities that produce changes in land cover, such as agriculture, mining, and urban development, are a major cause of the ecological footprint. Over the past century, the global human population has increased 3.5 times, while the area of agricultural land has doubled.[4] From 1980 the tropics have experienced higher deforestation than temperate regions, with the largest concentration in the Amazon Basin.[5] Changes in land cover such as deforestation affect the functioning of ecological systems at multiple scales, with consequences ranging from global climate change, soil and hydrological degradation, to increased biological extinctions.[6,7,8,9,10] Human pressures on the environment are expected to increase, at least in the near future, as a result of the expanding global population and increasing levels of consumption and waste accumulation.[3] Although the stakes are high, knowledge of the processes that underlie human-induced land cover change is still limited.[7] This is an emerging issue in spatially explicit environmental disciplines such as landscape ecology.[11,12]

The proximate causes of human-induced changes in land cover arise from both broad-scale clearing of natural vegetation for agriculture, mining and urban development, and from habitat modifications resulting from forestry and altered grazing and fire regimes. However, the ultimate causes are the biophysical and cultural factors that influence where and at what pace habitat clearing and modification occurs, fueled by the global population growth and the consumption of resources.[7] To reduce the human ecological footprint and encourage sustainable land use, we need to understand these ultimate causes.

Most human-induced land cover change is directly related to land use. Land use involves the exploitation of land and its resources and largely represents the interaction between humans and the biophysical environment.[1] Land cover change occurs at different spatial scales, including the local (10^2 to 10^3 km^2), regional (10^4 to 10^6 km^2), and national (10^6 to 10^8 km^2), or continental and global geographical scales. Land use is a process driven by social, economic, political, technological, and cultural factors, but is spatially and temporally constrained by biophysical conditions such as climate, soils, water availability, accessibility, and biological resources. These drivers and constraints, and their interactions, need to be conceptualized and analyzed in an integrated multidisciplinary framework if we are to develop an adequate understanding of the ways in which land use processes lead to land cover change. Bürgi et al.[12] point out that convincing integrative studies are still largely lacking, with several key areas requiring investigation: (i) moving from pattern-oriented to increasingly process-oriented studies, which include broad time frames and historical approaches; (ii) devising methods for extrapolation and generality building; (iii) finding ways to link data of different qualities; and (iv) effectively incorporating culture as a driving force of change. Global pressures of human-landscape interactions and their feedbacks have increased as a result of the world's sociopolitical, economic, and cultural globalization processes during the late 20th century. Because of this, opportunities and constraints for land use change, in particular new land uses, increasingly depend on the global factors influencing markets and policies at different organizational levels.

Historically, the major changes in global land cover occurred in temperate regions, but land cover change, especially deforestation, is now concentrated in tropical regions, leading to large reductions in net primary production.[8] It is widely recognized that deforestation rates in Latin America, especially in the tropical rain forests of the Amazon Basin, are currently the highest in the world.[6,13,14] About half of the land clearing in Latin America has taken place since 1960,[15] with the Amazon Basin accounting for a large proportion of recent global deforestation.[5] The most studied region of the Amazon is the Brazilian Amazon,[16,17,18,19,20,21,22,23,24] which shows that most of the land clearing has resulted from government-oriented colonization schemes and infrastructure building.

The Colombian Amazon region is less well studied, although it is currently a major global deforestation hot spot.[5,25] Unlike the Brazilian Amazon, clearing is occurring in an unplanned way, with peasants migrating freely into the region attracted by an illegal drug economy and the possibility to own land. Other regions of Colombia such as the Andes and the Caribbean are already heavily impacted by human land use.[26] In Colombia, a growing population and increasing integration into the global economy are bringing incremental pressures on natural ecosystems. Colombia is internationally recognized for its high biological diversity, both in terms of species richness and endemism.[27,28,29,30] However, knowledge of the processes and drivers of deforestation and consequent threats are limited.[28] There needs to be a much broader and deeper understanding of the spatial and temporal patterns of change and the factors driving these changes. This new knowledge will allow more meaningful assessments of the impacts of deforestation on ecosystems and biodiversity and provide useful inputs for land use and conservation planning.

In this chapter we present a multiscale synthesis to contribute to the understanding of the patterns, processes, and drivers of unplanned land cover change in the tropics, based on Colombia as a study case.[31] We address land cover change at the national, regional, and local levels, and analyze the variability of drivers, spatial patterns, and rates of change, focusing on the spatially dominant forest ecosystems. At the national level, attention is directed toward investigating the broad present and future land cover trends across the main biogeographic regions. At the regional level, we focus on the colonization of the Amazon region, while detailed analysis at the local level is carried out in six selected sample areas of lowland humid tropical forests. Finally, we discuss the implications of the research outcomes for conservation planning.

5.2 STUDY REGION

Colombia covers a land area of 1.1 million km^2, with large variations in altitude (0 to 5,800 m), mean annual rainfall (300 to 10,000 mm), length of growing period (60 to 360 days per year[1]), and a high diversity of geological substrates, giving rise to an unusual high environmental variability relative to its geographic size. Colombian ecosystems range from desert and tropical savannas to very humid rain forests and tropical snow-covered mountains. As a consequence of this variability, Colombia has high levels of endemism and species richness and has been classed as a megadiverse country.[28,32] The country can be divided into five biogeographic regions with specific biophysical and land use characteristics (Figure 5.1a): Andes (278,000 km^2), Caribbean (115,400 km^2), Pacific Coast (74,600 km^2), the Colombian Amazon (455,000 km^2), and Orinoco plains (169,200 km^2); and two smaller regions, the Magdalena (37,100 km^2) and Catatumbo (7,000 km^2), which are generally included in the Andean region.

The total population of Colombia is currently around 40 million, 25% of which still live in rural areas. The population has been historically concentrated in the Andean and Caribbean regions.[33,34] These regions currently have an average rural

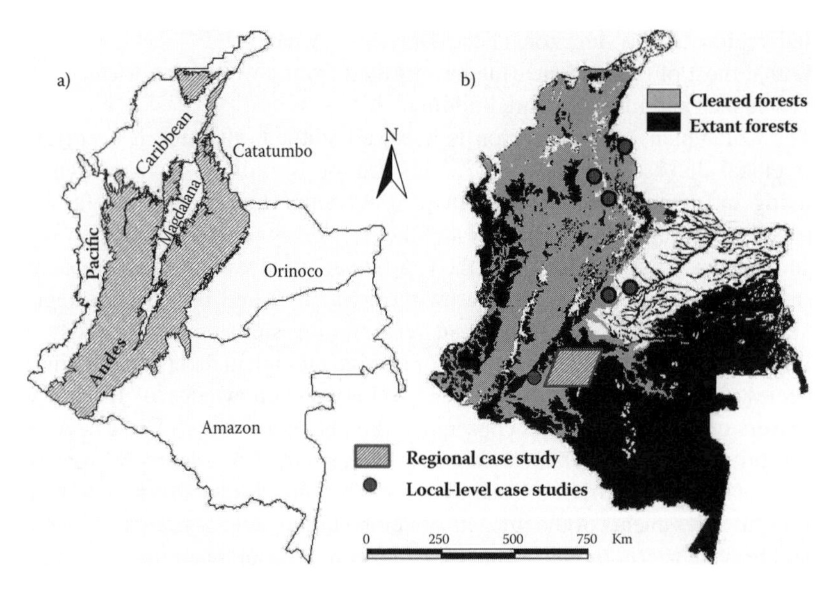

FIGURE 5.1 Location of Colombia showing: (a) the seven regions; and (b) extent of the original (gray + black) and remnant forested ecosystems (black), and nonforested ecosystems (white); and local and regional study areas (red).

population density of approximately 33 persons per km^2, while the Pacific, Orinoco, and Amazon regions have lower densities ranging from 5 to 17 persons per km^2. During the 20th century, the population grew 10-fold, having a strong impact on the landscape. However since the 1970s, Colombia has become an increasingly urban and industrialized country, with the national population growth rate falling below 2% in the late 1990s, and the rural population stabilizing. Although the country underwent a strong racial mixing resulting in a dominant Mestizo population, the cultural diversity is still high, with regionally contrasting rural cultures of 99 ethnic groups with 101 languages, varying from Amerindian to Afro-American and European-American.[35]

The economy is based on mining (oil, coal, and nickel), agriculture (coffee, flowers), and industrial exports. In recent decades, Colombia has experienced considerable social and political unrest, driven by extremist left- and right-wing armed forces, triggering large internal population movements and economic destabilization. Parallel to this unrest, a pervasive economy of illegal crops (coca and opium) for export has developed in many remote frontier areas, causing further social and political instability. These internal social and political pressures have important consequences for regional patterns of land cover change, including a redirection of colonization patterns and land abandonment in certain areas.

5.2.1 Regional Level

The regional study area (23,000 km^2) is located in the Caquetá Department in the Andean foothills of the eastern Amazon region of Colombia (Figure 5.1b). It includes 17 municipalities (whole or in part) with a total rural population of about 180,000 in 1993. Population densities range from 7 to 16 inhabitants per km^2 of cleared land. The

TABLE 5.1
General Biophysical Characteristics and Data Sets of the Local Study Sites

Study area	Area (km²)	Region	Dates of air-photo coverage	Mean altitude (m)	Mean annual rainfall (mm)/ No. dry months (<100 mm)	Original vegetation in landscape
La Balsa	128	Orinoco	1938, 1961, 1979, 1987, 1992, 2001	250	2500 / 4	Forest–Savanna mosaic
Guamal	140	Orinoco	1938, 1961, 1979, 1987, 1992, 1997, 2001	300	3000 / 3	Forest–Savanna mosaic
Caquetá	100	Amazon	1946, 1975, 1985, 1992, 2000	200	3100 / 2	Forest
Opón	68	Magdalena	1971, 1985, 1996, 2002	200	3200 / 1	Forest
Berrío	82	Magdalena	1950, 1961, 1977, 1985, 1996, 2002	120	3000 / 3	Forest
Tibú	74	Catatumbo	1961, 1975, 1985, 2000	150	2700 / 3	Forest

Source: Etter, A. et al.[40] With permission.

most extensive land use is cattle ranching, with illegal crops of coca (*Erythroxylum coca*) covering a smaller area but increasing in economic importance since the early 1980s. The latest figures are debated, but indicate a drop of almost 50% in illegal plantations area in Colombia from a high in 2001.[36]

5.2.2 Local Level

The local-level analysis focused on six landscapes located in the humid lowlands (> 2,000 mm mean annual rainfall) of the central (Magdalena) and eastern (Orinoco, Amazon, and Catatumbo) regions of Colombia. The landscapes comprise areas ranging from 68 to 128 km^2, which have been subject to substantial land clearing during the past 30 to 60 years (Figure 5.1b, Table 5.1). Less than two million people (5%) currently live in the rural lowlands. In these areas, the rural population density is very low, varying from less than 5 to 15 inhabitants/km^2, with lowest densities occurring in the extensive cattle grazing landscapes in the Caribbean, Orinoco, and Amazon regions.

5.3 METHODS

The data sets and methods described herein were used to analyze the spatial patterns of forest conversion, determine underlying drivers of this process, and predict forested areas with a high probability of conversion in the future. The study applied a multiple-scale modeling approach using logistic and ordinal regression, regression tree analysis, and geographical information systems (GIS). Land cover changes were modeled with biophysical and socioeconomic variables to analyze and predict deforestation and regeneration processes. For all levels of analysis, the response variable was a binary (1,0) map of forest cover, while the explanatory variables included soil fertility, rainfall, moisture availability index, slope, accessibility, and protected areas for the national level, and soil fertility, accessibility, and land cover neighborhood of forest and secondary vegetation for the regional and local analyses. We used a broad range of information sources, including remotely sensed data from aerial photographs and satellite images, and secondary sources of biophysical and socioeconomic data, as well as historical data (Table 5.2).

5.3.1 National Level

At the national level, the statistical modeling comprised two steps: (i) predicting the spatial location of forest conversion using logistic regression and regression trees; (ii) refining the predictions with the population growth rate data. We employed both modeling techniques as they treat the spatial variation in the effect of explanatory variables differently.[37] Logistic regression models the effect of variables in a spatially homogeneous manner, whereas classification trees can treat the effect in a spatially heterogeneous manner. Using the results of the best-performing model, we identified areas with a high probability of conversion as those with a predicted probability greater than 70%. We overlaid these areas with a recent rural population growth rate map to identify areas with a high probability of conversion and a population growth rate greater than 2%. We defined these areas as "deforestation hot spots." All analyses were performed using S-PLUS.[38] A qualitative assessment of the relative risk of forest conversion for the different remnant natural forest ecosystems was done by overlaying the predicted deforestation hot spots with the ecosystem map by Etter.[26]

5.3.2 Regional Level

Using the binary forest/nonforest maps, forest proportion zoning maps were produced by smoothing the data over a 50 by 50 grid cell window and then sliced into 10% intervals.[39] To analyze the spatial dynamics of deforestation and regeneration, we compared how deforestation and regeneration changed in each 10% forest cover zone for each time period. In a second step, we performed map cross-tabulations of the 10% zone maps for the three periods (1989 to 1996, 1996 to 1999, 1999 to 2002) (e.g., cells making transition from the 70% to 80% forest cover zone to the 60% to 70% forest cover zone in a given period of time). This allowed us to classify zone transitions in a spatially explicit way and to identify the intensity of change (negative for deforestation and positive for regeneration) depending on the number of zone jumps per period. We then defined hot spots of change only as those areas with "high speed" transitions, defined as areas showing two or more zone jumps (e.g., from the 30% to 40% zone to the 10% to 20% zone).

TABLE 5.2
Data and Data Sources for the Three Levels of Analysis

Level of analysis	Dependent variable		Independent variable	
	Type	Data source	Type	Data source
National	Binary forest/nonforest	Ecosystem map (Etter, 1998)	Climate, soil, slope, distance to roads, distance to rivers, rural population growth (1985–1993)	Instituto de Estudios Ambientales -IDEAM, 2004; Instituto Geográfico Agustín Codazzi, 1983, 1985, 2000; International Water Management Institute (IWMI), 2004)
Regional	Binary forest/nonforest	Landsat ETM (1989, 1996, 1999, 2002)	Soil, distance to roads, distance to rivers, rural population growth (1985–1993), neighboring land cover	Instituto de Estudios Ambientales -IDEAM, 2004; Instituto Geográfico Agustín Codazzi, 2000; Malagón et al., 1993)
Local	Binary forest/nonforest	Air-photos (1938–2000); Landsat ETM (2000, 2002)	Soil, distance to roads, distance to rivers, neighboring land cover	Instituto Geográfico Agustin Codazzi, 1983, 2000; Air-photos (1938–2000)

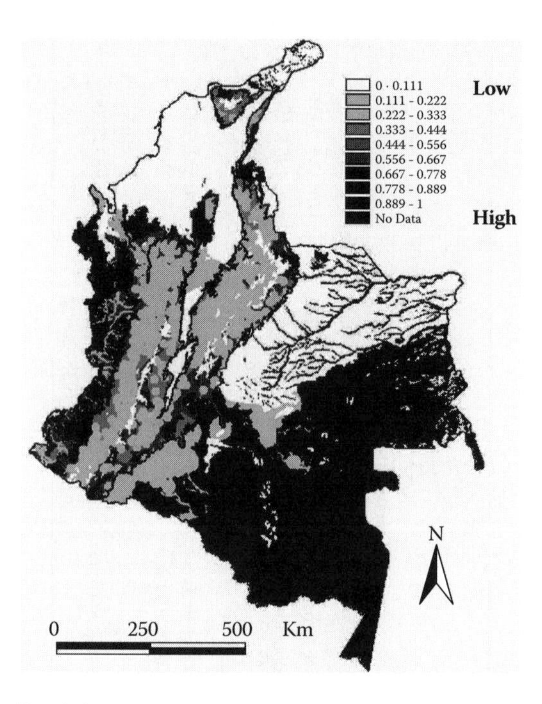

FIGURE 5.2 Predicted forest presence according to the best model (classification tree). (From Etter et al.[50] With permission.)

5.3.3 Local Level

Forest dynamics, as quantified through forest clearing and regrowth, were modeled and analyzed using five different logistic regression models, including an increasing number of independent variables, in order to explain the spatiotemporal patterns and some of the drivers of change.[40] The purpose of modeling was threefold: (i) to analyze the general temporal trends of the deforestation process of humid lowland forests in different regions; (ii) assess the relative importance of the predictor variables for each a priori model; and (iii) select a best approximating model from the a priori set of candidate models.

5.4 RESULTS

5.4.1 National Deforestation Threat Hot Spots and Main Drivers

The best performing model to predict deforestation was the regression tree that included the regions as a variable (Figure 5.2). At the national level, the most important variables explaining the presence and absence of forest cover were distance to roads and distance to towns for both the logistic regression and classification tree models. However, the effect of these variables on the presence of forest cover varied from region to region. The effect of other variables such as soil fertility (Andean, Pacific, Orinoco) and number of rain days (Caribbean, Magdalena) are important explanatory variables in only a few regions. In general, the following relationships

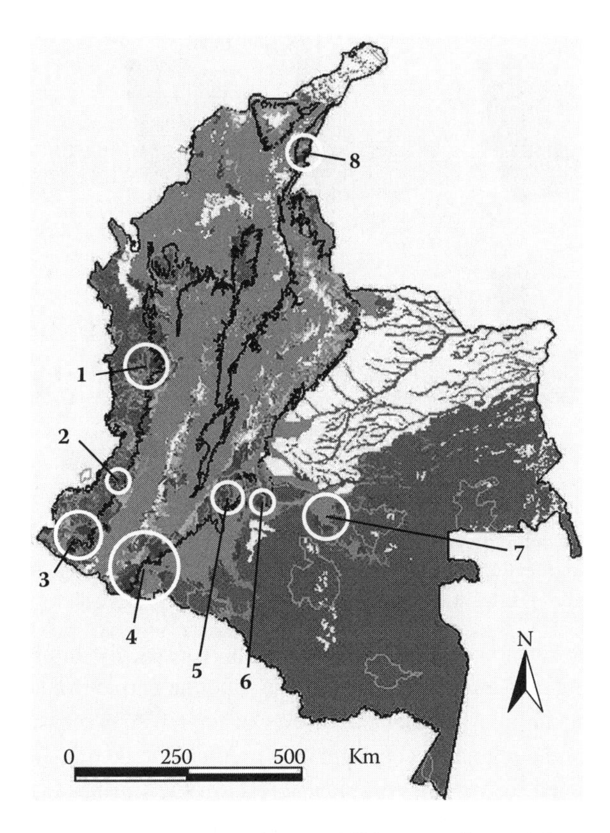

FIGURE 5.3 (See color insert following p. 132.) Predicted deforestation hot spots obtained by combining areas predicted to have the highest probability of forest conversion (>70%) from the best model (the region-specific classification tree) with the areas with greater than 2% rural population growth rate (1985–1993). Red depicts the deforestation hot spots (areas with >70% probability of forest conversion and >2% rural population growth). Orange and red depict areas with >70% probability of forest conversion. Green depicts forested areas, gray represents cleared forested areas, and white represents nonforested areas. White circled areas indicate current hot spots of deforestation, which are also areas of high-value biodiversity value: (1) Quibdó-Tribugá, (2) Farallones-Micay, (3) Patía-Mira, (4) Fragua-Patascoy, (5) Alto Duda-Guayabero, (6) Macarena, (7) Guaviare, and (8) Perijá. Black line is the Andean region, and light green lines are national parks. (From Etter et al.[50] With permission.)

between deforestation and the explanatory variables were observed across all models: deforestation was predicted to be greater in unprotected areas that have fertile soils, gentle slopes, and are near to settlements, roads, and rivers. The relationship between the number of rain days and forest conversion was positive in the Caribbean and Amazon, while negative in the Andean region. The regression tree model adjusted with the population growth data shows a set of distinct geographical "hot spots" of predicted future deforestation areas (Figure 5.3).

The five forested ecosystems predicted to be most vulnerable to forest conversion according to the predicted future deforestation hot spots ranked according to predicted deforestation area were: the humid tropical forests of the undulating plains

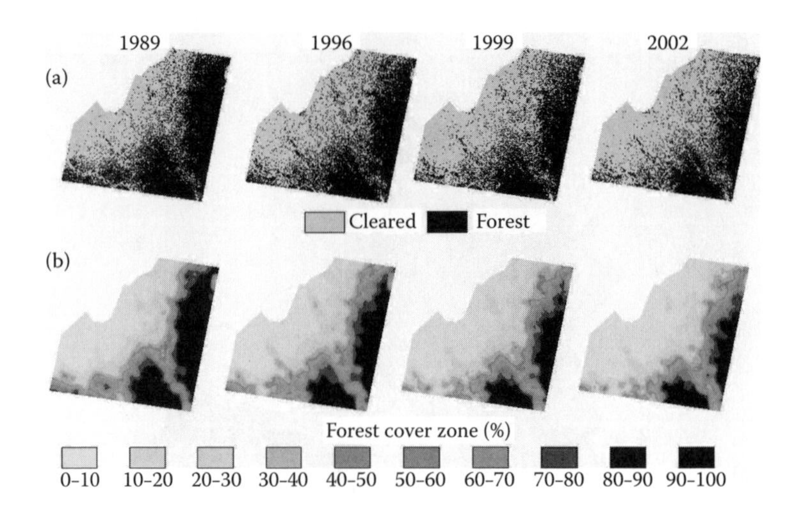

FIGURE 5.4 (See color insert following p. 132.) Forest maps of the colonization front for each study date: (a) extent of forest cover (black = forest); and (b) percentage forest cover at 10% increment zones. (From Etter et al.[39] With permission.)

of the northern Amazon, the humid sub-Andean forests, the humid high-Andean forests, the humid mid-altitude Andean forests, and the humid tropical forests of the undulating plains in the Magdalena. However, when risk was measured in terms of the proportion of the remnant ecosystem area predicted to be transformed, the ranking of risk changed to: very dry tropical forests in the Caribbean, humid tropical forests of the rolling landscapes in the Magdalena, the tropical dry forests of the hills, and the lowland swamp forests of the Caribbean.

5.4.2 Amazon Colonization Fronts

The analyses show the colonization front in the Amazon moving like a wave originating in the main urban centers of the region and following primarily the rivers (Figure 5.4). Between 1989 and 2002, the 90% forest cover boundary, used as a proxy for the "colonization frontline," moved to the east an average distance of 11.2 km, representing a rate of 0.84 km per year. However, the rate varied between 0.5 to 3.2 km per year depending on the location along the front. The total annual net deforestation rates showed a peak of 40,400 ha in 1996 to 1999, increasing from 18,600 ha in 1989 to 1996, and declining to 23,830 ha in 1999 to 2002. An average of 25,400 ha of forests was cleared each year during 1989 to 2002, of which 3,400 ha per year was cleared in the perforation process beyond the 90% frontline.

The net forest change rate within the colonization front was the result of the combined effect of deforestation and forest regeneration and varied with the forest cover zones. The highest rate of deforestation occurred in the 50% to 80% forest cover zones, with rates exceeding 4% per year during 1996 to 1999. The forest regeneration rate peaked in the 20% to 50% forest cover zones, with annual rates of up to 0.9%, suggesting the progressive increase of secondary forests against

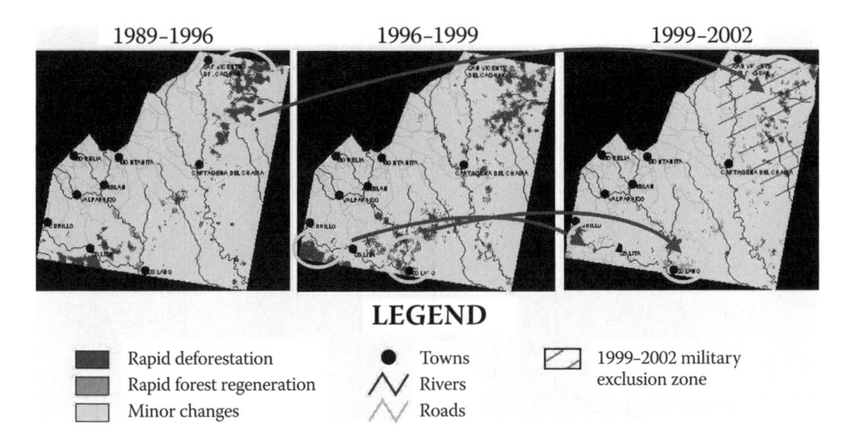

FIGURE 5.5 **(See color insert following p. 132.)** Spatial location of the local hot spots of deforestation (red) and regeneration (green) for the three time periods of study. (From Etter et al.[39] With permission.)

"mature forests" as the level of landscape transformation increases. The same pattern was maintained for all time periods.

High rates of forest cover change were observed in localized areas, or "local hot spots" of forest loss and forest regeneration (Figure 5.5). Some of these deforestation hot spots were spatially stable during the entire study period, such as the one in the northeast of the study area; others advanced; while some locations showed intense deforestation in one period followed by a period of intense regeneration, such as the southwestern part of the study area. The more dynamic landscapes were invariably situated in the forest cover zones with an intermediate amount of remnant forests (30% to 70%). The northeast of the study area, which was declared in December 1999 as a military exclusion zone by the government for the peace process with the FARC guerrillas, showed the most rapid deforestation in the 1999 to 2002 period, while the rest of the region showed the opposite trend.

5.4.3 Deforestation Patterns at the Local Level

The analysis of the deforestation trend in all of the six lowland forest areas during the past 60 years showed a rapid decline of the forest cover, and was typically followed by an increase of pasturelands, initially supporting extensive cattle grazing, but intensifying where infrastructure development permits. However, a major finding was that in all cases the forest loss follows an asymmetrical logistic pattern, with four recognizable phases of forest loss (Figure 5.6):

Phase i: initial phase with low rate of change
Phase ii: middle phase with highest rate of change
Phase iii: mid to late phase with rate of change slowing down
Phase iv: final phase with an apparent new dynamic equilibrium of re-growth balancing forest loss

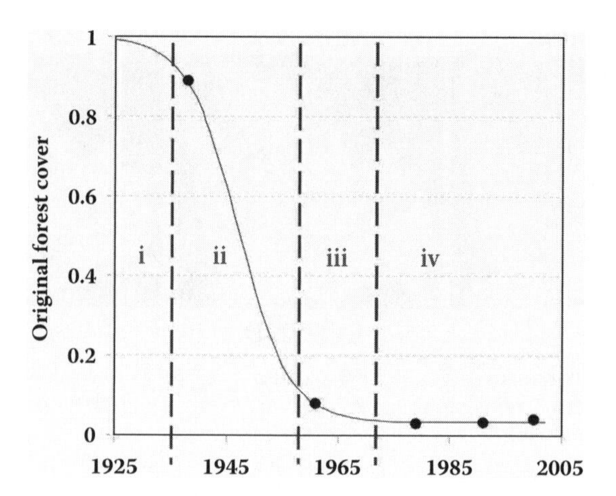

FIGURE 5.6 Logistic pattern of forest cover decline during the transformation process in Colombia, showing phases i to iv. Large plot corresponds to La Balsa case study, with equation: Y = 0.033 + [0.967 / (1 + exp(0.219 * (X – 1947)))]. (From Etter, A. et al.[40] With permission.)

An additional result is that as a consequence of the logistic decline pattern of the forest cover in the landscape, the rates of deforestation show a quadratic relationship with the proportion of forest in the landscape (Figure 5.7a) and a direct linear relationship with the amount of exposed forest edge (Figure 5.7b). The maximum rates of clearing occurred at intermediate values of forest proportion in the landscape and at the highest values of exposed forest edge densities, of approximately 40 m ha^{-1}.

Also, when comparing the various regions, differences in colonization waves and rates of deforestation were evident for this period in the Amazon, Orinoco, and Magdalena Caribbean. The logistic curve depicting the deforestation process shifts in time depending on when the land clearing process began. On average, the time span between beginning (phase i) and end (phase iv) of the process is attained after 30 to 40 years, at a time when the remnants of forest ecosystems reach values between 2% and 10% of their original cover. In all studied landscapes, the dominant and more persistent replacement cover was introduced pastures, with crops representing a minor proportion of land use following clearing.

5.5 DISCUSSION AND CONCLUSIONS

Our study contributes to broadening the understanding of patterns and drivers of tropical deforestation processes in unplanned colonization fronts. There are four major contributions about patterns and processes of land cover change in Colombia that can be drawn from this study with potential application in other tropical regions.

First, loss of forests follows a logistic pattern. The case studies from lowland forests in Colombia show that the decline of forest cover during the process of deforestation conforms to a logistic pattern, reaching a semistable transformed landscape state after undergoing four transformation phases with varying rates of deforestation.

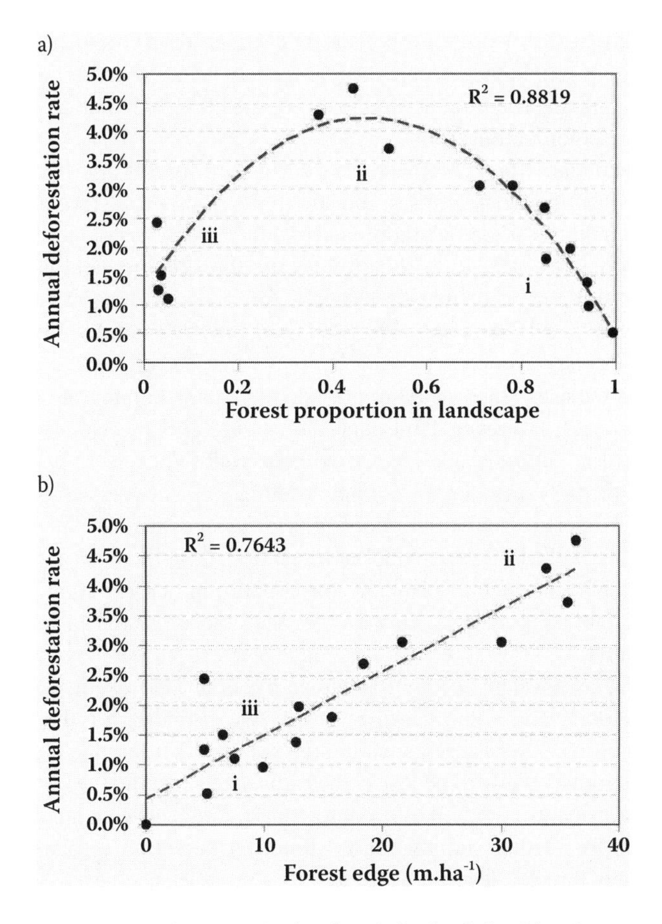

FIGURE 5.7 Forest cover change at the local scale in the Colombian Amazon: (a) Relation between deforestation rates and forest proportion in landscape; (b) Relation between rates of deforestation and forest edge density. (From Etter et al.[51] With permission.)

This means that the rate of forest loss assumes a quadratic shape with a peak rate of loss when approximately half the landscape is transformed.

Second, deforestation fronts move as waves. In the unplanned colonization fronts in Colombia, deforestation progresses in a relatively uniform pattern at an approximate speed of 1 km per year, mimicking a wave. Forest cover declines within the advancing wave, following a similar logistic pattern of forest decline mentioned above, thereby linking the local and regional levels of analysis.

Third, changes in landscape structure, such as an increase in edge density per unit area, facilitate the clearing process. There is a direct relationship between the exposed forest edge density in the landscape and the rate of deforestation, with maximum rates of clearing occurring at the highest forest edge densities of approximately 40 m ha^{-1}. The relationship between proportion of forest in the landscape and forest edge assumes a quadratic form, with maximum edge density at intermediate forest proportions when the clearing rate is also highest. This suggests a strong link between spatial pattern (forest proportion and edge density) and process (rate of forest clearing) in unplanned tropical colonization fronts. Furthermore, this relationship also confirms the broader applicability of the logistic pattern of forest loss across an entire colonization front.

Fourth, secondary forests replace cleared mature forests and become a dominant component of the tropical forest mosaic in the highly transformed landscapes. The historical analysis of deforestation across all case studies demonstrates that in highly transformed tropical forest landscapes the forest component is made of two parts: a stable component of mature remnant forest and a dynamic component of secondary forests of different ages. These components are spatially mixed, and form a successional forest mosaic.

Landscape metrics[1,41] are meant to provide the means to integrate issues linking pattern and process. However, landscape indices are still mostly used to describe the spatial patterns of landscape change without making a link to processes in the landscape.[42] Our study provides an example where a link between landscape metrics, such as forest proportion and amount of exposed forest edges, and the speed of land conversion is made. The logistic model of forest decline at the base of these findings provides an important general principle for explaining deforestation in unplanned colonization fronts (Figure 5.8). However, it needs to be tested in tropical forests outside Colombia.

Studies and assessments on multiscale analyses of land use models have been done for Ecuador[43] and Central America.[44,45] We provide, for the first time in Colombia, a multiple-level overview of deforestation, which should give both general (i.e., where are forests likely to be lost in the future) and specific (i.e., how and why is deforestation advancing) guidelines for detecting spatially explicit land cover change dynamics both for further studies and to direct the attention of planners. We use specific data sets for each level of analysis.

Land use planning in Colombia is limited with regard to the use of up-to-date and relevant spatial data, which is critical considering that Colombia is a country where the biological natural resources are an important asset and the gains and losses from inappropriate land cover change are very high. Understanding the level of threat to natural ecosystems is fundamental to help make more informed decisions in conservation planning.[46] The general patterns of deforestation described make a significant contribution to this aim and can potentially be applied in other tropical and subtropical countries of the world where unplanned land clearing occurs. The study has specific relevance to land use and conservation planning in highly dynamic landscapes because the models help predict and anticipate the spatial progression of unplanned deforestation fronts. The results should, in principle, be transferable and applicable to other unplanned colonization fronts in the tropics and subtropics and help improve land use and conservation planning by: (i) providing methods to calculate the threat to forest ecosystems and characterize the dynamics of clearing associated with colonization fronts; (ii) assisting the analysis of carbon sequestration and greenhouse gas emissions budgets,[47,48] through the spatial and temporal dynamics of deforestation and forest regeneration; (iii) guiding biodiversity conservation planning and restoration ecology in highly transformed and dynamic lowland forest landscapes, thereby helping discriminate between rapidly changing (hot spots) and more stable areas.[49]

FIGURE 5.8 (a) Andean landscape showing high levels of transformation with dominating cattle grazing land uses. (b) High diversity forests in the Amazon region of Colombia. (c) Colonist forest clearings in an Amazon colonization front in Colombia.

We showed that an unplanned tropical forest landscape transformation results in severely cleared and impacted landscapes. The logistic model of forest decline provides an important general principle for explaining deforestation in unplanned colonization fronts. However, it needs to be tested in tropical forests such as other regions of Latin America, Africa, and Asia. However, an important question remains: does government controlled land-use planning make a difference? An intriguing research agenda would be to investigate the extent to which these land cover change patterns and processes occur in countries where land clearing is planned, such as Brazil, or partially regulated, such as Australia.

ACKNOWLEDGMENTS

We especially thank Hugh Possingham, Kerrie Wilson, Stuart Phinn, and David Pullar for their input at several stages of this research. Universidad Javeriana and the University of Queensland provided financial support to AE.

REFERENCES

1. Forman, R. T. T. *Land Mosaics: The Ecology of Landscapes and Regions.* Cambridge University Press, New York, 1995.
2. Wackernagel, M. et al. Tracking the ecological overshoot of the human economy. *PNAS* 99(14), 9266, 2002.
3. Vitousek, P. M. et al. Human domination of Earth's ecosystems. *Science* 277(5325), 494, 1997.
4. Houghton, R. A. The worldwide extent of land-use change: In the last few centuries, and particularly in the last several decades, effects of land-use change have become global. *Bioscience* 44(5), 305, 1994.
5. Lepers, E. et al. A synthesis of information on rapid land-cover change for the period 1981–2000. *Bioscience* 55(2), 115, 2005.
6. Laurance, W. F. Reflections on the tropical deforestation crisis. *Biological Conservation* 91, 109, 1999.
7. Lambin, E. F. et al. The causes of landuse and land-cover change: Moving beyond the myths. *Global Environmental Change* 11, 261, 2001.
8. DeFries, R., and Bounoua L. Consequences of land use change for ecosystem services: A future unlike the past. *GeoJournal* 61, 345, 2004.
9. Fearnside, P. M. Global warming and tropical land-use change: Greenhouse gas emissions from biomass burning, decomposition and soils in forest conversion, shifting cultivation and secondary vegetation. *Climate Change* 46(1–2), 115, 2000.
10. Myers, N. et al. Biodiversity hotspots for conservation priorities. *Nature* 403, 853, 2000.
11. Wu, J., and Hobbs R. Key issues and research priorities in landscape ecology: An idiosyncratic synthesis. *Landscape Ecology* 17, 355, 2002.
12. Bürgi, M., Hersperger, A. M., and Schneeberger, N. Driving forces of landscape change—Current and new directions. *Landscape Ecology* 19(8), 857, 2004.
13. Bilsborrow, R. E., and Ogendo, O. H. W. O. Population-driven changes in land use in developing countries. *Ambio* 21(1), 37, 1992.
14. Achard, F. et al. Determination of deforestation rates of the world's humid tropical forests. *Science* 297(5583), 999, 2002.

15. Houghton, R. A., Skole, D. L., and Lefkowitz, D. S. Changes in the landscape of Latin America between 1850 and 1985 II. Net release of CO2 to the atmosphere. *Ecology Management* 38(3–4), 173, 1991.
16. Frohn, R. C. et al. Using satellite remote sensing analysis to evaluate a socio-economic and ecological model of deforestation in Rondonia, Brazil. *International Journal of Remote Sensing* 17(16), 3233, 1996.
17. Fearnside, P. M. Soybean cultivation as a threat to the environment in Brazil. *Environmental Conservation* 28(1), 23, 2001.
18. Moran, E. F. et al. Effects of soil fertility and land-use on forest succession in Amazonia. *Forest Ecology and Management* 139(1–3), 93, 2000.
19. Portela, R., and Rademacher, I. A dynamic model of patterns of deforestation and their effect on the ability of the Brazilian Amazonia to provide ecosystem services. *Ecological Modelling*143(1–2), 115, 2001.
20. Skole, D. L. et al. Physical and human dimensions of deforestation in Amazonia. *Bioscience* 44(5), 314, 1994.
21. Alves, D. S. Space-time dynamics of deforestation in Brazilian Amazônia. *International Journal of Remote Sensing* 23(14), 2903, 2002.
22. Laurance, W. F. et al. Predictors of deforestation in the Brazilian Amazon. *Journal of Biogeography* 29, 737, 2002.
23. Laurance, W. F., and Ferreira, L. V. Effects of forest fragmentation on mortality and damage of selected trees in Central Amazonia. *Conservation Biology* 11(3), 797, 1997.
24. Laurance, W. F. et al. The future of the Brazilian Amazon. *Science* 291(5503), 438, 2001.
25. Viña, A., Echavarria, F., and Rundquist, D. C. Satellite change detection analysis of deforestation rates and patterns along the Colombia-Ecuador border. *Ambio* 33(3), 118, 2004.
26. Etter, A. Mapa General de Ecosistemas de Colombia (1:2 000 000). In: Chaves, M. E. and Arango, N., eds., *Informe Nacional sobre el Estado de la Biodiversidad en Colombia—1997.* Instituto Alexander von Humboldt, Bogotá, 1998.
27. Myers, N. Biodiversity hotspots revisited. *Bioscience* 53(10), 916, 2003.
28. Chaves, M. E., and Arango, N., eds. *Informe Nacional sobre el estado de la Biodiversidad en Colombia—1997.* I. A. von Humboldt, Bogotá, 689, 1998.
29. Convention of Biological Diversity (CBD). Global Biodiversity Outlook, 2005. Available from: http://www.biodiv.org/gbo/.
30. Orme, C. D. L. et al. Global hotspots of species richness are not congruent with endemism or threat. *Nature* 436(7053), 1016, 2005.
31. Etter, A. Modeling unplanned landscape change: A Colombian case study. Dissertation, University of Queensland, Brisbane, 2006.
32. Hernández, J. et al. Estado de la biodiversidad en Colombia. In: Halffter, G., ed., *La Diversidad Biológica de Iberoamérica*, Vol. I. CYTED, Mexico, 1992.
33. Herrera, M. Ordenamiento Territorial y Control Politico en las Llanuras del Caribe y en los Andes Centrales Neogranadinos, Siglo XVIII. Dissertation, University of Syracuse, New York, 2000.
34. Colmenares, G. *Historia Economica y Social de Colombia*. 5th ed. Vol. 1, TM Editores, Bogotá, 357 pp, 1999.
35. Loh, J., and Harmon, D. A global index of biocultural diversity. *Ecological Indicators* 5(3), 231, 2005.
36. United Nations Office for Drug Control UNDOC. Colombia: Coca cultivation survey. United Nations Office for Drug Control and Government of Colombia, Bogotá, 72 pp, 2004.
37. McDonald, R. I., and Urban, D. L. Spatially varying rules of landscape change: lessons from a case study. *Landscape and Urban Planning* 74(1), 7, 2004.

38. Insightful-Corporation. S-Plus 6.1 for Windows. Seattle, Washington, 2002.
39. Etter, A. et al. Characterizing a tropical deforestation front: A dynamic spatial analysis of a deforestation hotspot in the Colombian Amazon. *Global Change Biology* 12, 1409–1420, 2006.
40. Etter, A. et al. Modeling the conversion of Colombian lowland ecosystems since 1940: drivers, patterns and rates. *Journal of Environmental Management* 79(1), 74–87, 2006.
41. McGarigal, K. et al. FRAGSTATS: Spatial pattern analysis program for categorical maps. Computer software program produced by the authors at the University of Massachusetts, Amherst. Available at: www.umass.edu/landeco/research/fragstats/fragstats.html, 2002.
42. Li, H., and Wu, J. Use and misuse of landscape indices. *Landscape Ecology* 19(4), 389, 2004.
43. de Koning, G. H. J. et al. Multi-scale modelling of land use change dynamics in Ecuador. *Agricultural Systems* 61, 77, 1999.
44. Kok, K., and Veldkamp, A. Evaluating impact of spatial scales on land use pattern analysis in Central America. *Agriculture, Ecosystems and Environment* 85(1–3), 205, 2001.
45. Kok, K. et al. A method and application of multi-scale validation in spatial land use models. *Agriculture, Ecosystems and Environment* 85(1–3), 223, 2001.
46. Wilson, K. et al. Measuring and incorporating vulnerability into conservation planning. *Environmental Management* 35(5), 527, 2005.
47. Houghton, R. A. Aboveground forest biomass and the global carbon balance. *Global Change Biology* 11(6), 945, 2005.
48. Foley, J. A. et al. Global consequences of land use. *Science* 309(5734), 570, 2005.
49. Meir, E., Andelman, S., and Possingham, H. Does conservation planning matter in a dynamic and uncertain world? *Ecology Letters* 7(8), 615, 2004.
50. Etter, A. et al. Regional patterns of agricultural land use and deforestation in Columbia. *Agricultural Ecosystems and Environment* 114, 369–386, 2006.
51. Etter, A. et al. Unplanned land clearing of Columbian rainforests: Spreading like disease? *Landscape and Urban Planning* 77, 240–254, 2006.

6 Landscape Dynamism

Disentangling Thematic versus Structural Change in Northeast Thailand

Kelley A. Crews

CONTENTS

6.1 INTRODUCTION

Land change research necessarily draws upon an interdisciplinary milieu of theories and practices ranging from ecology to geography to policy and beyond; a dominant approach successfully used in this arena over the past few decades has been that of *scale-pattern-process*.[1] Choice of scale influences which landscape patterns can be discerned, in turn used to infer process. The number of resulting landscape studies have increased substantially over the past decade.[2,3,4,5] Assessing sensitivities of pattern detection and subsequent inferable processes to changes in scale (typically spatial resolution or pixel size) of remotely sensed data has become an important research agenda for remote sensing specialists.[6,7] This work draws in particular on principles of landscape ecology that posit the possible impacts that scale can have on landscape characterization.[8] Scale is comprised of two primary facets: grain, the size of an observational unit (e.g., the dimensions of a single pixel), and extent, typically represented as the size of the overall study area. Although these component parts are typically applied to spatial scale, they as easily may be applied to temporal scale (e.g., scale as the frequency of observation and extent as the total length of study).

Implicit in these arguments is the separation of landscape configuration from landscape composition. In other words, the spatial association of different elements is as important as the overall proportion of the landscape occupied by the elements. Many land use/land cover change (LULCC) applications, ranging from biology conservation to hydrological assessments to land use planning, now routinely provide this decoupled information.[4] This work reviews the successes, limitations, and possibilities of enriching LULCC research with increased temporal grain or observational frequency for extricating compositional or thematic change from configuration or structural change. A case study from Northeast Thailand is used to illustrate this paired approach, underscoring the need to further develop and refine this method in ecosystems from elsewhere on the naturally or anthropogenically driven spectrum and with varying degrees of spatiotemporal heterogeneity.

6.1.1 Temporal Frequency: Tensions and Limits

Although most scale-pattern-process work has focused on spatial scale, temporal scale has nonetheless been explicitly included in theoretical discussions even if seldom analyzed.[4,9] Environmental remote sensing defines scale more broadly to include spatial, temporal, spectral, radiometric, and directional scales.[10] Spatial and temporal scale are particularly important when extracting thematic data for satellite image-based change detection.[11] Spatial scale, both grain and extent, is regarded as a major influence on detection and definition of landscape patterns.[12,13,14] Temporal frequency, or the time scale between available data acquisitions, is less studied in LULCC work, in large part due to the limited availability of high quality and high resolution multi-temporal image sequences.[6,15,16] The temporal grain of imagery, though typically not referred to as such, has been examined in environments where seasonality (whether due to phenological, climatic, or anthropogenic changes) can interfere with assessment of longer-term (read: interannual) LULCC.[17,18,19,20] The temporal extent of LULCC projects typically defaults to either the early 1970s (concomitant with the 1972 launch of Earth Resource Technology Satellite or ERTS 1, later renamed Landsat 1) or, in a few cases, to a few decades earlier when military reconnaissance aerial photography was available. The necessarily truncated temporal extent of these studies presents problems in establishing baselines, a critical issue given the necessity of determining what change has occurred and placing it in the appropriate historical context.[21,22] The term spatiotemporal scale or domain is widely used within the LULCC modeling community, but this description may cause some confusion. The coinage of the term presents an understandable commitment to consider how landscapes change across time and space, though currently the state of the science tends to model spatial interactions over time rather than offering a path for digitally representing a spatiotemporal scale as interactive rather than only combinatory.

Compounding the confusion is the dialogue concerning the use of the terms *landscape scale* and *landscape level*. For geographers, the term landscape scale connotes a certain size (spatial extent) of a study area—larger than a plot, smaller than a continent,[4] and the term level is often used to denote study across spatial scales,[23] whether or not such work is spatially explicit (e.g., multilevel modeling[24]). For many ecologists who tend to focus first on biotic components of landscapes (e.g., populations and communities), the term landscape scale is nonsensical, since depending upon organism size and range, a landscape can be incredibly small (consider microbes living in a small puddle) or as large as a planet.[25] This tension manifests itself in incongruities and inconsistencies in the use of these terms (as well as whether they are seen as interchangeable or not), a troublesome glitch for LULCC scholars drawing upon ecology through the lens of landscape ecology.[1,26] For the purposes of discussing LULCC in this work, the term *landscape scale* will be used to connote spatial and temporal grain and extent commonly used in LULCC work. The term *landscape level* will be used to refer to organizational or theoretical constructs where the landscape lies on a spectrum of functional units, ranging from patches to landscapes to metalandscapes.[27]

Note the above issues revolve primarily around spatial scale; rarely do LULCC practitioners mention a landscape scale when referring to a certain time, as opposed to those studying longer-term landscapes (e.g., in geomorphology, sedimentology, or palynology). Temporal matters are receiving more scholarly attention of late, particularly in both empirical and process-based modeling efforts.[28] Landscapes in temporally shallow LULCC studies are being increasingly considered as acting upon their previous incarnations[22] and seen, therefore, as temporally contingent upon those past drivers. Path dependencies can and are being trained into modeling scenarios, and presumably other temporal analogues of spatial concepts will be operationalized (e.g., spatial neighborhood effects could be used as a model for more sophisticated representations of path dependency via temporal neighborhood effects). In spatial neighborhood effects, it is understood that the precise location of the neighbor relative to the area of interest is often unimportant as long as that neighbor is within a certain thresholded distance. So while analysis may take place in a spatially explicit environment, conditions or rules can be written to loosen that explicitness and query, for example, for neighbors within a certain spatial distance (without regard to that distance), without regard to direction. The parallel in temporal studies would be to relax the assertion of temporal explicitness such that it may not matter when a particular preceding event happened, only that it did happen or how often it happened. Although temporal modeling is fairly straightforward in terms of assessing causality (given the presumption of linear time moving only forward), the challenge remains to sort out from myriad periodicities of landscape drivers and change, which are important enough in any given landscape, and how then to best define a temporal landscape scale.

6.1.2 From Pattern and Structure to Process and Function

Several decades of LULCC research have shown that understanding landscape change requires detecting changes in both composition and configuration.[6,8] Typically these components are assessed sequentially: first, a landscape is classified into a thematic land use/land cover (LULC) scheme for at least two times; second, the configuration of each of those classified landscapes are quantified through some type of pattern assessment (often pattern metrics[29]); and third, postclassification change detection is performed on those thematic classifications to produce thematic change map(s).[30] Although the process usually stops there, some researchers have also then quantified the configuration of the change map(s) with pattern metrics as well,[11] though concerns of error propagation have limited this approach.[31] The importance of ascertaining spatial structure (and changes in said spatial structure) stems from landscape ecology, where spatial configuration facilitates and mitigates the flow of energy and materials across the landscape.[8] That is, the landscape interactions that both cause and are manifested as landscape change necessarily occur in space, and location matters. That is not to say all changes occur as diffusive movements since, depending upon the vector of movement, energy or materials may be imparted by jumping or percolating across the landscape.[4] Defining the temporal nature of spatial structure will assist in taking these measurements and converting the scale-pattern into process.

Process can be defined in two primary ways. The first will be referred to as dynamics, and it is a mechanistic concept rooted in patterns of change, growth, and activity; this definition is embraced by the Geographic Information Science community (GISc) studying landscape dynamics, and fits the necessarily piecemeal fashion by which LULC is extracted, studied, and modeled. The second type of process will be referred to as dynamism, which is a more gestaltic concept that involves continuous change, growth, or activity; this definition comes from the ecology community (particularly landscape ecology) and fits the more continual nature of the processes studied by ecologists, whether particular to landscape studies or not.[32] The nuance of the difference in these two approaches is slight, but the implications are easily observable in the varying operationalization of both epistemology and methodology now evidenced in landscape change studies from these two communities. Here, panel analysis of LULC and paneled pattern metrics are offered as one method of bridging this gap, suggesting that LULCC scholars shift toward an approach of understanding landscape dynamism via improved assessment of LULCC dynamics.[11,27] That is, improved description of mechanics should lead to improved explanation and prediction of process and, perhaps, function.

6.2 A PANEL APPROACH

Panel analysis simply refers to a longitudinal method whereby units of analysis are held constant. Long used in psychology and sociology, panel or longitudinal analysis followed the same subjects over time (as opposed to a census or cross-sectional approach, where different subjects are evaluated in each observation period). Technically, all from-to remote sensing-based change detection is panel analysis, since each pixel or instantaneous field of view (IFOV) is followed individually through time, presuming accurate geometric rectification.[30] However, from-to change detection usually is performed on pairs of images, whereas panel analysis (in LULCC) is now used to refer to a time series of three or more classifications.[33,34,35] In panel analysis, pixel histories or trajectories are constructed that maintain the entire temporal pattern of LULC in order to reveal greater information about process(es) behind observable patterns. For example, consider a humid tropical area classified only into forest (F) and nonforest (N) and observed over two decades every other year. With panel analysis, trajectories that might be calculated would include those suggesting semipermanent deforestation (e.g., F-F-F-F-N-N-N-N-N-N-N), deforestation and successional regrowth (e.g., F-F-F-F-N-N-N-N-N-F-F), afforestation or reforestation (e.g., N-N-N-N-N-N-N-F-F-F-F), silviculture of fast growing tree species (e.g., F-N-N-N-F-N-N-N-F-N-N-N), or fallow cycling (e.g., N-F-F-F-F-F-N-F-F-F-F-F-N). With traditional from-to change detection of the first and last years, those trajectories would have had their change characterized as follows: semipermanent deforestation with F-N would still be called deforestation (correct); deforestation and successional regrowth with F-F would be called stable or permanent forest (incorrect); afforestation or reforestation with N-F would still be called as such (correct); silviculture with F-N would be called deforestation (incorrect), and fallow cycling with N-N would be called permanent nonforest (incorrect). Ultimately the panel approach to LULCC does nothing to improve attribution of classes that are stable over time, and little to improve attribution of classes whose change is unidirectional. But landscape components that undergo very quick change, cycle through multiple stages, switch between two or more classes frequently, or are influenced by relatively short-term phenomena (e.g., seasonality) are open to better multitemporal characterization. That is, panel analysis improves our ability to detect the kinds of change that LULCC research is largely designed to capture, model, and manage; by corollary, traditional from-to change detection is biased toward detecting stable, slow-changing, or unidirectionally changing classes. As the number of classifications in the time series increases, quite obviously the ability to detect greater nuanced or more quickly switching change increases. The question for LULCC projects then is how many images are enough? The textbook answer is that it depends upon the time footprint of landscape processes on the landscape (e.g., humid tropical forests reach successional canopy closure more quickly than the average temperate forest); the practical answer is that it depends on how many quality images are available in an area given atmospheric interference, sensor problems, cost of acquisition, and access to archives, to name only a few of the problems facing the LULCC community.

6.2.1 Extension to Pattern Metrics

Though pattern metric analysis is typically output as statistics at the patch, class, and landscape levels, some packages such as Fragstats[29] allow for outputting patch-based images whereby each patch (from which all patch, class, and landscape statistics are generated) is mapped with a unique identifier or object (whether computed in raster or vector, bit depth limitations notwithstanding). The goal of paneled pattern metric analysis is to assess the changing structure of landscape patches without regard to thematic class. That is, in building pattern metric panels we explicitly choose to examine the nature of, for example, fragmentation without regard to whether it is an urban area, forested expanse, or agricultural field that is being fragmented. By doing so, the explicit contribution of configuration as opposed to composition can be tested, assessed, and modeled. Current research at this point has focused on the formulation and sensitivity analyses of paneled pattern metrics, and this method requires further testing in other ecosystems and landscapes with differing levels of spatial, temporal, and spatiotemporal heterogeneity. In cases where the robustness and sensitivity of the paneled pattern metric method is validated, the next step is to not only test the separate impacts of composition and configuration, but also their interaction and confounding as well.

The construction of paneled pattern metrics follows logically from panel analysis, and the entire panel method is presented in Figure 6.1. First, a time series of imagery is categorized into thematic classifications; a minimum of four temporal observations is suggested, though if patch boundaries can be derived, generated, or found elsewhere, three classifications may suffice (in absence of preexisting patch delineations, a baseline year of the time series is used, requiring three further classifications for moving beyond traditional two-image change detection). From these classifications, a panel LULC is created as depicted and as described above. Additionally, pattern metric analysis is run on each classification, outputting both statistics and patch images for all observations for each metric of interest. For purposes of this discussion, presume the metric of interest is the interspersion/juxtaposition index (IJI). Change images between consecutive pairs of patch images are calculated and may initially be left as float output but must eventually be binned into categories of change (e.g., increase by > 20%, increase by 10% to 20%, increase by 5% to 10%, change by ± 5%, decrease by 5% to 10%, decrease by 10% to 20%, decrease by > 20%). Once binned appropriately, the change between each set of IJI metric images is stacked to build a trajectory of change at the patch level and then exported to individual pixels and built back to a final mapped product of paneled pattern metrics output at the patch level.[11] The process is repeated for each metric of interest, with each metric binned according to appropriate hypothesized or observed thresholds or flip points.

Currently bounded or constrained metrics have been tested in order to limit the subjectivity involved in categorization of the metric output. That is, metrics such as IJI, double log fractal dimension, and percentage landscape all—as operationalized in Fragstats and other pattern metric programs—have theoretical bounds where both the upper and lower limits are known. Unbounded or unconstrained metrics (e.g., mean patch size, shown in Figure 6.1 for contrast) present greater subjectivity in categorization since there is no theoretical limit for these metrics (though in any given landscape and with any given classification scheme an empirical limit obviously exists). As currently written, the paneled pattern metric algorithm presumes equal intervals between time steps since the original time series used for testing met those conditions; modification to account for differing time lags is easily done via a weighting mechanism once categorization thresholds (number and placement) have been determined. As such, the method is suitable for both interannual and intraannual analyses.

6.3 THAI TESTING GROUNDS

The concern over interannual and intraannual LULCC stems from building this approach in an environment with strong phenological, climatic, and anthropogenic seasonal pulses, rendering assessing longer-term LULCC problematic when anything but anniversary date imagery was used for deriving LULC information. Northeast Thailand is home to a region known as Isaan, where the former Nang Rong district resides (due to growth and redistricting this area now includes not only the Nang Rong district but also Non Suwan—denoted on some maps as Nong u Wuan, Chamni, and

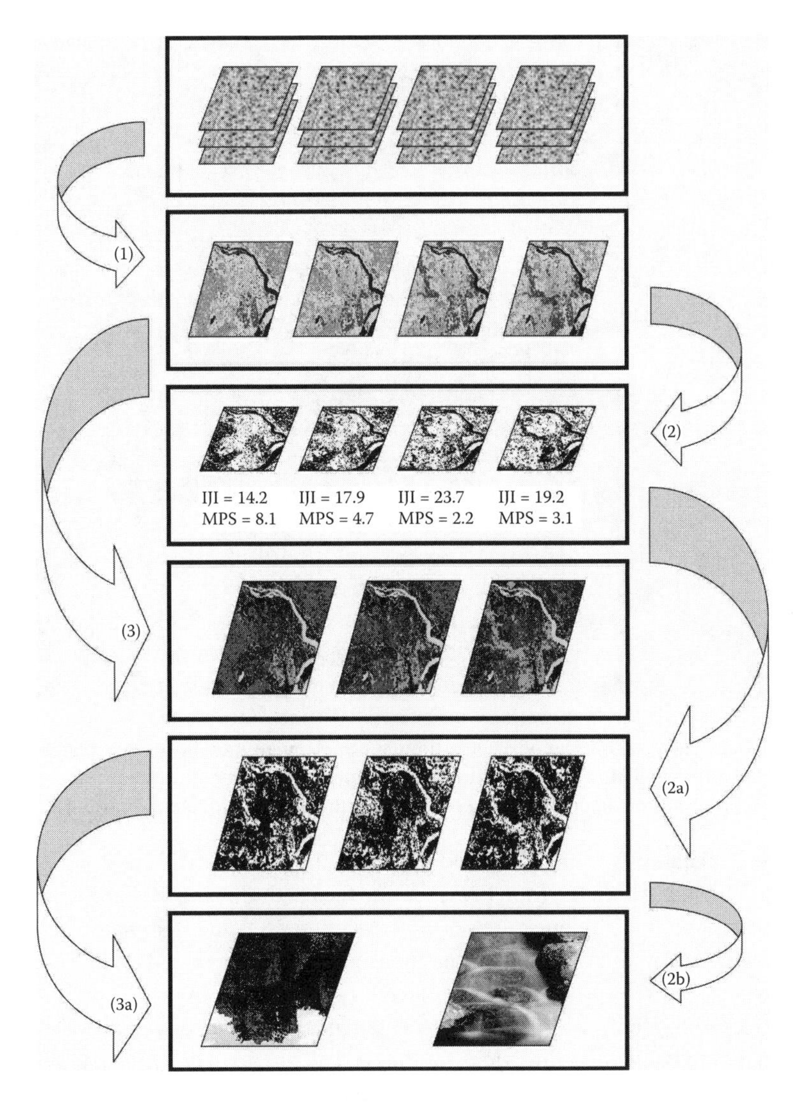

FIGURE 6.1 **(See color insert following p. 132.)** The panel process, conducted at both the pixel and patch levels: (1) four multispectral satellite images are each categorized into a thematic LULC classification; (2) pattern metrics are run on each of the four LULC classifications, each producing a set of patch, class, and landscape statistics (here the interspersion/juxtaposition index [IJI] and mean patch size [MPS] are shown) as well as an output image of the delineated patches; (2a) pattern metric output for each of the four times is used to calculate three piecemeal change maps for each pattern metric and each consecutive pair of images (e.g., showing fluctuations in IJI or MPS between two time periods) as per Crews-Meyer[11,27]; (2b) three pattern change maps are stacked into one panel of all structural change for each given metric (e.g., showing fluctuation in IJI or MPS through all time periods) as per Crews-Meyer[11,27]; (3) three thematic change maps are created for each of the time periods represented by the four classifications; (3a) the three thematic change maps are stacked to represent the full record of all thematic change across the four classifications as per Crews-Meyer.[3]

FIGURE 6.2 Typical nuclear village settlement as seen in 1:50,000 scale panchromatic aerial photo from 1994, with approximate settlement boundary indicated. Note remnant forest patches used for shade relief, and rice paddy surrounding village radially.

Chalerm Prakeat; this work was tested primarily in current day Nang Rong and Non Suwan). Situated in both Buriram Province and the north-flowing Mekong River Delta system, the area is the poorest area of a poor country[36,37] and dominated culturally, ecologically, and financially by a strong monsoonal pulse, poor soils,[38] and concomitant lowland wet rice production.[39] Villagers typically live in a nuclear settlement pattern (see Figure 6.2), with residences located in lowland wooded remnants and rice fields radiating out in most directions for the typical 2 to 5 km daily walk to fields.[40,41] Though this area was not influenced by the Green Revolution, agriculture has driven the conversion of the landscape initially opened by military road building efforts and facilitated by the gradual building toward a market economy.[37] Wet rice replaced savanna in the lowlands, while drought-deciduous crops such as cassava and sugarcane followed the 1970s factor price increase into the upland dry dipterocarp forests. Following a currency collapse in the late 1990s, many young adults who typically migrated to Bangkok or the eastern seaboard for labor returned to the district at the same time the government underwent a new wave of decentralization across federal to local levels.[40] An increasingly dense network of road building and water impoundments,[33] combined with poor environmental management (e.g., lack of draining rice irrigation waters increases soil salinity), has compounded the intensification cycle seen in parts of Southeast Asia and elsewhere. Although these longer-term dynamics have been documented through an extensive household and community survey series

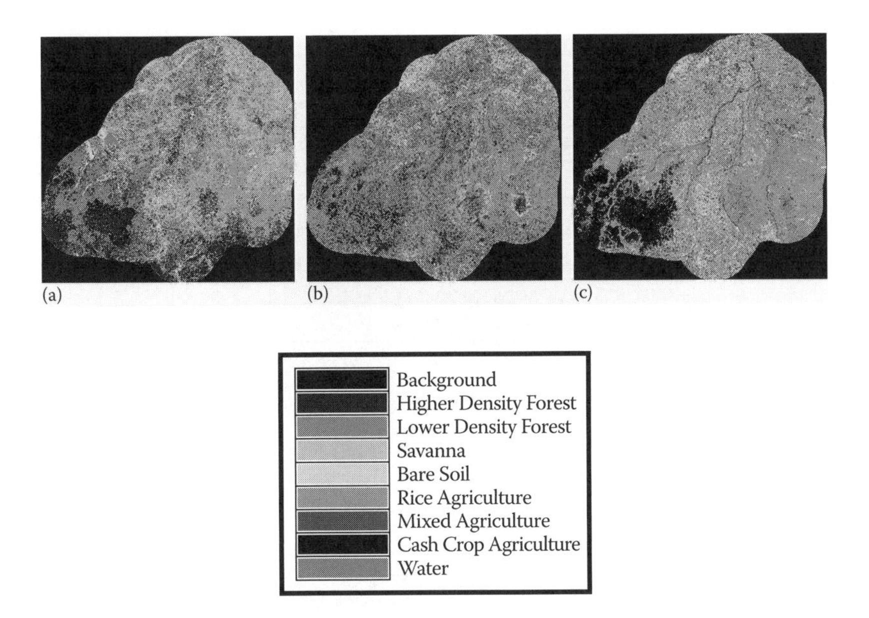

FIGURE 6.3 **(See color insert following p. 132.)** (a) LULC in the greater study area in the 1972/1973 water year; (b) 1985; and (c) 1997.

as well as remote sensing and geographical information systems (GIS) analyses, the seasonal pulses also detected (when imagery, fieldwork, and weather permit) can cause detectable landscape change as large in magnitude (if not ecological importance) as two decades of interannual change.[18,19] The presence of a monsoonally driven climate adds to the logistical problems of obtaining cloud-free imagery for deriving LULC information. However, a deep time series has been established as part of a larger project and has proven more than adequate for testing the panel LULC and paneled pattern metric methods.[41,42] Figure 6.3 illustrates interannual trends in LULCC in the larger study area over a 25-year period; easily discernible are the rapid decline in more highly vegetated LULC (particularly in the upland southwestern section) and the expansion of rice into the lowland savannas.

6.3.1 Local Lessons Learned Thus Far

Figure 6.4 illustrates a stylized representation of four LULC classes and their compositional change over time as observed and/or reported elsewhere. Figure 6.4a shows the interannual or longer-term change in forest (primarily upland dry dipterocarp and gallery remnant forests along riparian corridors), savanna (primarily lowland graminoids with some standing trees), wet rice agriculture, and other agriculture (upland or drought deciduous crops and cash crops, including cassava, kenaf, jute, and sugarcane).[33] These "real" changes can be contrasted with the stylized representation of intraannual change in a given year due to previously mentioned seasonality shown in Figure 6.4b. This graph is ordered by the Thai water year that runs April 1 through March 31, with early monsoonal showers (known as mango rains) commencing in May and followed by several months of heavy precipitation that is extremely

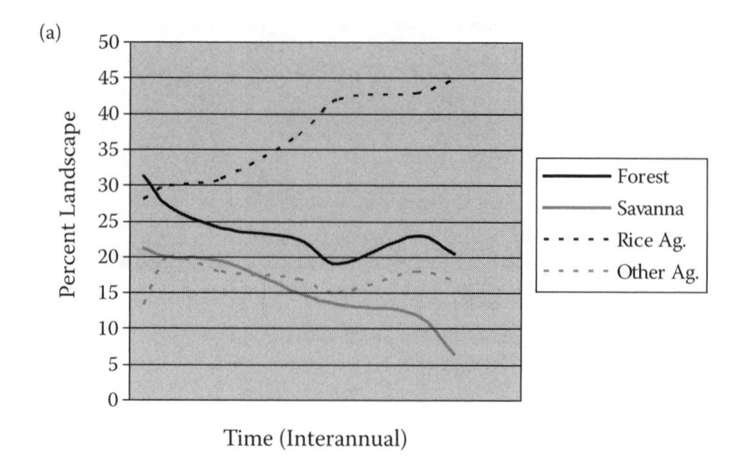

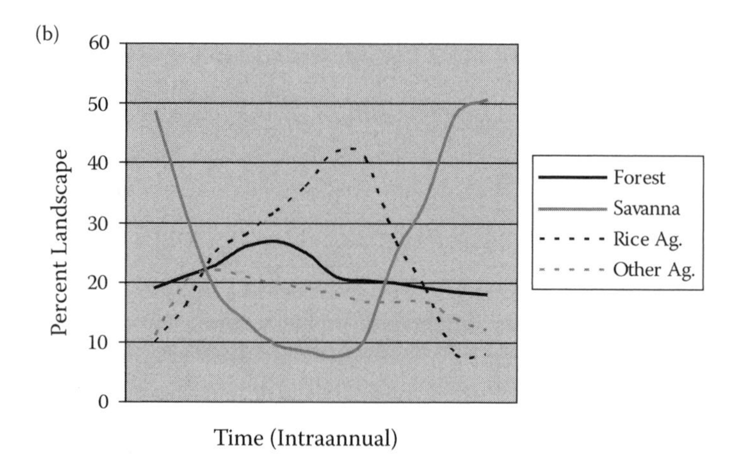

FIGURE 6.4 (a) Stylized LULC trends observed and/or reported in Northeast Thailand from the 1970s to late 1990s (annual change, holding seasonality constant). (b) The same trends within a given typical year (intraannual).

variable in both time and space; rice is typically harvested in late November or December, with fields burned usually in January and the driest months ending the water year. These "changes" are part real (e.g., phenological change with agricultural crops or deciduous cycles) and in part artifact (e.g., green-up from showers without actual canopy or biomass change).

Typical forest changes in this part of northeast Thailand represent a familiar story: from the 1970s through the 1990s, forests generally declined (as did savannas) due to agricultural extensification. An early rise in other agriculture in the uplands at the expense of forests (now relegated to extremely thin riparian corridors and small remnants atop the most upland sites on volcanic soils) was followed by a sharp rise in wet rice agriculture in the lowland areas. Village settlement and expansion occur in these lowland areas as well, although these areas account for little change in terms of spatial footprint. Some tradeoff is suggested between wet rice agriculture and other agriculture in dry versus wet years, likely explained by terrace position (not anthropogenic rice terraces, but fluvial terraces or middle elevation grounds that do not flood each year) since rice may be planted at slightly higher elevations in wet years with cash crops occupying those areas in dryer years. Overall the interannual compositional changes exhibit some sharp increases or decreases over time but represent fairly stable trajectories (note this graph is overall composition; a per pixel comparison would represent much more switching among classes, and a greater classification scheme depth or move away from an anniversary date/stylized representation would show much greater fluctuation as well).

The intraannual representation depicts much more marked fluctuations in composition of these four LULC classes. Forests appear to have greater cover due to phenological changes (green-up from rains) and lesser cover during deciduous events. Rice agriculture changes dramatically with crop calendar and related monsoonal timing, as rice paddy move from flooded to planted to flowering to harvest to burning to barren; so too do lands of other agriculture change, although to a lesser extent than rice owing to the drought resistant nature of some of these crops. Savanna cover appears to change dramatically as well and does so in response not only to green-up of grasses but moreover to all agricultural lands spectrally mixing with grassy savannas during dryer periods. The magnitude of intraannual change compared to interannual change underscores the need for anniversary date imagery, but also for understanding the process implications of the periodicities of different types of change. Without a physical process guide in terms of the criticality of a certain loss or gain in one cover type over another, perhaps LULCC scholars can at least bracket interannual change in the context of seasonal dynamics typically observed to understand when the landscape has changed beyond its "natural" resilience.[9,43]

Figure 6.5 moves to the consideration of structural change by showing two typical metric trajectories for interannual changes, again from the 1970s to the 1990s. Figure 6.5a graphs the IJI for the same four classes, and from this illustration the landscape narrative quickly becomes apparent. A mapped view of this for a subset of the study area is also presented in Figure 6.6, as typically in interpretation both patch-derived statistics (e.g., Figure 6.5) and maps of change in configuration (Figure 6.6) are used. Here, both forest cover and wet rice agriculture experienced a sharp increase in IJI followed by a sharp decrease, indicating increases in interspersion followed by decreases. Taken together, these two trends exhibit evidence of an important ecological change in the landscape, and one that when mapped shows elevational differences. Forested areas, notably in the southwest in Non Suwan district, had been the matrix or dominant class in upland areas in the early 1970s, but became increasingly fragmented and interspersed as other agriculture was introduced. By the early 1990s, the forest had been desiccated to little but remnant patches, and although still interspersed with other LULC types, these forested patches were so small that the metric plummets as less and less forest edge remains to neighbor other LULC types. A similar trend of rice agriculture occurs in the lowland areas but for the opposite reason: rice experiences a sharp increase in interspersion as it became more and more widespread, until by the late 1980s it became so ubiquitous that its spatial cohesion results in lower IJI scores. Smaller changes in other agriculture IJI

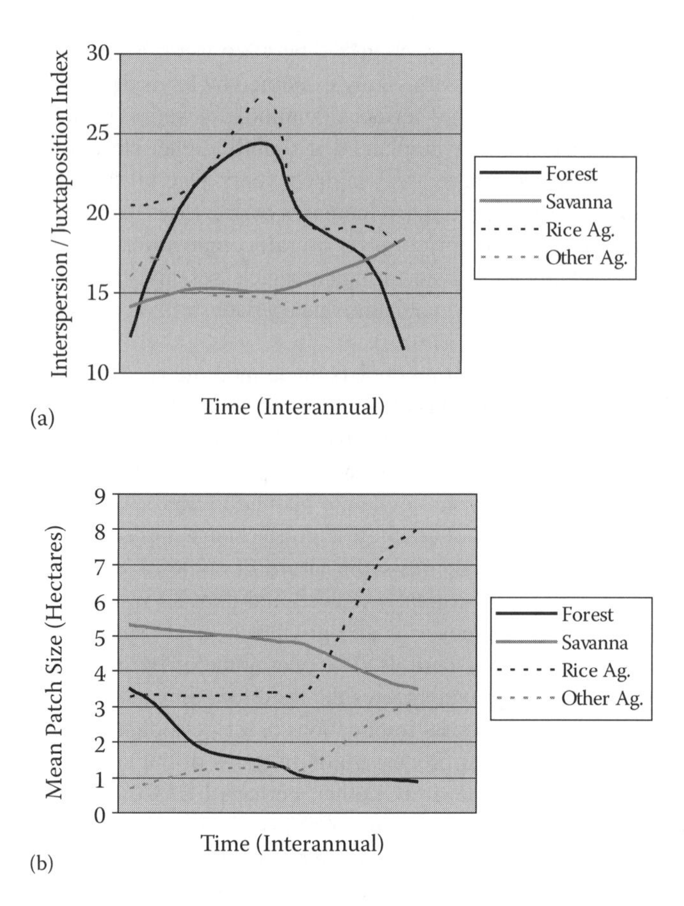

FIGURE 6.5 (a) Stylized LULC pattern metric change for the interspersion/juxtaposition index (IJI) observed and/or reported in northeast Thailand from the 1970s to late 1990s (annual change, holding seasonality constant). (b) Metric output for mean patch size (MPS) for the same time and location.

support the hypothesis that not only the composition but also the configuration of this class changes in the middle elevation areas, while experiencing some changes in upland areas at the expenses of forested lands. Savanna, particularly in the eastern portion of the area, becomes increasingly interspersed with wet rice agriculture, particularly in areas more proximate to the primary river channels (i.e., that flood the most frequently). This landscape narrative is bolstered by interestingly parallel changes in mean patch size (MPS) (Figure 6.5b), where typical agriculture patches (regardless of type or topographic position) increase as agriculture overwhelms the landscape at the spatial expense of both forest and savanna.

Moving beyond the combined changes in composition and configuration to configuration only[11,27] illustrates trends in structure related to topography and accessibility not detectable in the above efforts. First, waves of fragmentation are easily discernible in the southwestern or upland areas as people literally moved up the hills from surrounding low-lying areas. Typically upland crops are less labor intensive,

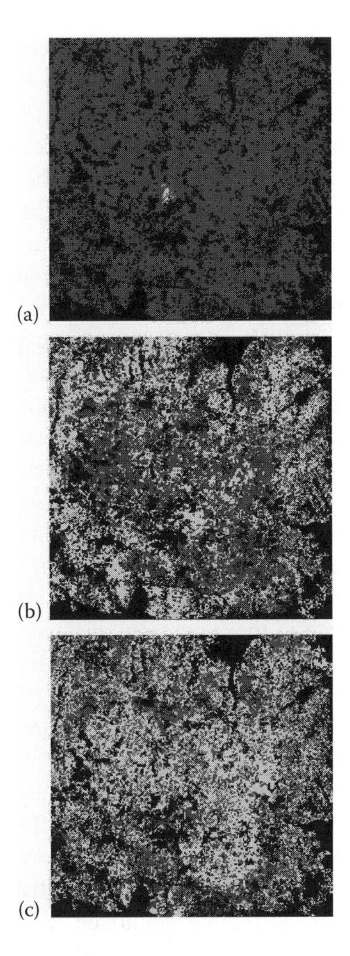

FIGURE 6.6 **(See color insert following p. 132.)** (a) The change in configuration from 1972/1973 to 1975/1976, revealing that the entire subset area has experienced a greater than 10% increase in interspersion/juxtaposition index (IJI) scores (due to increased fragmentation and concomitant interdigitation). Note that most of the area experienced the same type of change. (b) Illustration of a different trend between 1975/1976 and 1979, whereby increases, decreases, and relative stability in IJI vary spatially. More upland areas (most central in the subset) experienced a consolidation on the landscape, while peripheral areas remain relatively stable in terms of configuration with notable exceptions on the southeastern perimeter. (c) Illustration of the continued spatial heterogeneity in IJI, with lowland/peripheral areas undergoing continued fragmentation, while the less accessible, upland areas appear to have leveled off in terms of larger-scale fragmentation or consolidation but continue to experience small pockets of fragmentation throughout.

and do not require proximity to the nuclear village settlements that rice paddy do. Thus, when factor prices for cassava increased in response to European demand for cattle feed, it was relatively easy for lowland villagers to plant upland areas not previously claimed given that they had been seen as undesirable for rice or residence, given their (uphill) distance to rivers. Interestingly, the southern edge of this area did not see an advancement of agriculture as did the other areas, possibly due to that side's proximity to the Khmer (Cambodian) border and related history of military violence and mine fields.[40] In the lowland areas, the trends over time respond more to accessibility to riparian corridors and thus assurance of yearly flooding for the best rice paddy lands.[27] Structurally, though, the upland and lowland areas were very similar, despite the different dominant LULC types and various drivers of LULCC. First, in terms of IJI, the core areas (either most central in the upland areas or closest to riparian channels) exhibited the greatest possible fluctuation in IJI across time. Regardless of LULC type, these areas experienced increases, decreases, and periodic episodes of stability in IJI over the 1970s and 1980s. So while the cover types may have been relatively slow to change over time, by comparison the structure, specifically the interspersion of LULC classes, was constantly changing and changing in different ways. That is, not only was there first order change but also second order change. The areas proximate to these core regions also experience fairly dramatic structural change as reflected by IJI, although slightly more consistent in the second order (roughly two periods of increasing IJI with one period of decrease). In terms of percentage landscape (PCT), the similarities in upland versus lowland areas were less pronounced. The core upland areas (the least accessible) saw early decreases in PCT, followed by increases and subsequent decreases, similar to the trends in LULCC and LULC pattern change discussed earlier. The regions proximate to the upland core but more accessible (lower elevations) showed two periods of decrease as well, although the increase in PCT came at slightly different times (a notable exception was the southernmost region mentioned earlier, which was structurally stable in the dynamic sense in that it experienced a decrease in PCT through all observations). Lowland areas, in contrast, nearly uniformly displayed the decrease-increase-decrease pattern for PCT, despite observed stability in LULC composition, suggesting that even in the face of relatively stable LULCC the structural dynamism of the area was in "constant fluctuation."

6.4 PANELED PATTERN METRICS: MEANS OR END?

The subjectivity of the panel approach, and particularly the paneled pattern metric method, presents significant challenges and merits testing in ecosystems with different levels of human impact and landscape heterogeneity (across time and space). A particular concern is the delineation of patch boundaries; when testing existing patches (be they forest refugia, control plots, or cadastre-defined parcels of land), the boundaries to use in analysis are clear. But otherwise, boundaries are constructed from a year of the imagery, and the determination of the base year impacts what trends are possible to detect and in what direction. By testing the sensitivity and robustness of results to changes in the base year, researchers have in paneled pattern metrics a potential tool for creating ecologically meaningful units of analysis that fit within a hierarchical (nested or non-nested) framework suitable for drawing upon theories of landscape ecology and hierarchical patch dynamics.[4,8,9,44] With increasing grain in the time series (greater number of observations or images), the possible bias in baseline determination should drop. However, traditional notions of accuracy assessment calculation posit that as the number of temporal observations increases, the amount of accuracy assessment data needed increases at an increasing rate[31] and the likelihood of an acceptable cumulative product decreases sharply. If 10 images with 95% accuracy were used, the traditional manner of representing change accuracy would mean that, at best, the 10-image change product would have an accuracy of just under 60% (0.95^{10}): an accuracy level hardly worth pursuing. And yet it appears that these multi-input products offer the most promise for extracting process. Perhaps a new framework for considering how to think about change and accuracy is needed. From physics, Griffiths's[45] consistent history approach holds promise as a metaphor for accuracy, where the greater the observations, the less important any given error in any given observation is, since the pattern is indicative of larger temporal relationships at work and not a product of any one singular event or mistaken pixel classification. Rectifying this line of reasoning with disturbance and (dis)equilibrium theory presents a challenging but fruitful avenue of epistemological and methodological reasoning, especially given continued sensor systems failures requiring analysts to build time series from multiple sensor platforms and thus possibly introducing artifacts into classifications when compared across time.[30,31]

Integrating not only classifications derived from more than one sensor system but other kinds of products presents another frontier of LULCC work. Due to factors varying from cost to climatic conditions, establishing sufficient temporal grain or frequency challenges many research teams; moreover, the temporal extent of most LULCC work lacks the historical depth necessary to test, assess, or account for landscape impacts occurring over longer time frames or in the more distant past: geomorphological change, species evolution, past civilization land uses, and climatic change research all offer evidence that the forces at work on landscapes thousands to hundreds of thousands of years ago still impact landscapes and landscape components today (Figure 6.7).[22] Linking of mapped products to recent LULC classifications has been done infrequently to extend a time series, but the real challenge lies in linking non–wall-to-wall or nonspatially explicit products to today's digital products. Travel journals, agricultural taxation records, and artwork offer rare glimpses into unchartered temporal extents of landscapes of long ago,[22] even in those lacking in spatial extent and explicitness.

The panel method generally is, perhaps obviously, of most use in data sets with a rich time series. As digital archives of LULC and LULCC become more widely available, this approach can be further tested in differently impacted social and environmental landscapes. Panel analysis has most commonly been applied to information extracted from optical sensor systems, but it could be extended to other imagery sources. Panel analysis offers particular appeal for landscapes with temporal heterogeneity. Paneled pattern metrics, more specifically, offer the greatest potential insight when landscape cover change appears to be separable from changes in landscape configuration. For example, in areas of shifting or swidden (an area cleared for temporary cultivation by cutting and burning the vegetation) cultivation, the amount

FIGURE 6.7 (a) Seasonality impacts this landscape in several ways. Here, rice stubble is a typical "winter" or dry season landscape component. Haze in background is smoke from traditional burning of the rice fields to boost nutrients in these degraded soils. (b) Cash crop agriculture tends to occur in topographic regions not suitable for annual rice production. Shown are sugarcane (for market) and eucalyptus (typically for local building supplies, field borders, or soil stability). (c) Typical recently improved (elevated) road with rice agriculture in background. "Borrow pits" provide road elevation materials and are often subsequently used as small water impoundments for irrigation. (d) While most water impoundments in the area are small in spatial extent, the impact of the network of such impoundments is critical in population-environment impacts. Water impoundments, while used for irrigation and flood control throughout many fields, are typically ringed by a small fringe of water-demanding crops for household consumption, such as watermelon and bananas.

of lands under any particular use may change very little between observations; but the configuration could still be fluctuating, particularly under localized management strategies. Landscapes undergoing even drastic LULCC may in fact be structurally stable when wholesale replacement of classes occurs. Landscapes undergoing both compositional and configurational change present the most complex situation for landscape assessment, and disentangling these types of landscape change is critical for extracting a better understanding of process and function from pattern. Temporally, landscapes with heterogeneous change (different rates of change over different periods) would benefit more from the general pattern approach than those with consistent temporal processes, but proper extraction of these patterns presumes an adequately rich time series.

The panel method generally, as used here and elsewhere,[34] offers LULCC researchers one way of sliding closer to process on the pattern-process spectrum. Assessing dynamism (continuous change) from dynamics (changes and drivers assessed via snapshots in time) remains a critical area of concentration for the LULCC community. The paneled pattern metric approach provides a means for exploring stronger linkages to process and function from patterns of LULC and LULCC. Moreover, paneled pattern metrics constitute an explicit test of the value of landscape ecology to LULCC work by exploring the relative contributions and interactions of landscape composition and configuration. The implementation of this tool is, currently, prone to subjectivity. Although pattern metrics have been found to reveal critical differences in landscapes, rarely are they explicitly and quantitatively linked to human or biophysical processes. As such, determining appropriate thresholds for categorization of constrained or unconstrained metrics remains critical to do but difficult to justify. As with the introduction of vegetation indices several decades ago, a statistical correlation may convince some of the utility of an approach, but ultimately it is the empirical and quantitative tie to process that convinces practitioners of the approach's worth. The pursuit of that linkage is a ripe area for research of ecologically and biophysically grounded LULCC teams. It may be that the process linkage between paneled pattern metrics and landscape processes is never discovered or verified, or even that it is rejected and disproven. But until such a time, requiring a process linkage may be premature, when the greatest promise of paneled pattern metrics may lie in application of a data mining approach to first uncover critical thresholds or flip points of LULCC, and then bring to bear theories and methods for fleshing out the processes at work.

ACKNOWLEDGMENTS

I am grateful to the following for support of this work: Carolina Population Center and Department of Geography, University of North Carolina; Sigma Xi Scientific Research Society; and Mahidol University, Thailand.

REFERENCES

1. Walsh, S. J., and Crews-Meyer, K. A. *Linking People, Place, and Policy: A GIScience Approach*. Kluwer Academic Publishers, Boston, 2002.

2. Meyer, W. B., and Turner, B. L., II, eds. *Changes in Land Use and Land Cover: A Global Perspective.* Cambridge University Press, Cambridge, 1994.
3. Liverman, D. et al. *People and Pixels: Linking Remote Sensing and Social Science.* National Academy Press, Washington, D.C., 1998.
4. Turner, M. G., Gardner, R. H., and O'Neill, R. V. *Landscape Ecology in Theory and Practice: Pattern and Process.* Springer, New York, 2001.
5. National Research Council. *Population, Land Use, and Environment: Research Directions.* Panel on New Research on Population and the Environment. Entwisle, B., and Stern, P. C., eds., Committee on the Human Dimensions of Global Change, Division of the Behavioral and Social Sciences and Education. National Academies Press, Washington, D.C., 2005.
6. Frohn, R. C. *Remote Sensing for Landscape Ecology: New Metric Indicators for Monitoring, Modeling, and Assessment of Ecosystems.* Lewis Publishers, Boca Raton, Fla., 1998.
7. Rindfuss, R. R. et al. Developing a science of land change: Challenges and methodological issues. *Proceedings of the National Academy of Science* 101(39), 13976–13981, 2004.
8. Forman, R. T. T. *Land Mosaics: The Ecology of Landscapes and Regions.* Cambridge University Press, Cambridge, 1995.
9. Wu, J., and Loucks, O. L. From balance of nature to hierarchical patch dynamics: A paradigm shift in ecology. *Quarterly Review of Biology* 70(4), 439–466, 2005.
10. Jensen, J. R. *Remote Sensing of the Environment: An Earth Resource Perspective.* Prentice Hall, Upper Saddle River, N.J., 2000.
11. Crews-Meyer, K. A. Characterizing landscape dynamism via paneled-pattern metrics. *Photogrammetric Engineering and Remote Sensing* 68, 1031–1040, 2002.
12. Ahl, V., and Allen, T. F. H. *Hierarchy Theory: A Vision, Vocabulary, and Epistemology.* Columbia University Press, New York, 1996.
13. Gustafson, E. J. Quantifying landscape spatial pattern: What is state of the art? *Ecosystems* 1, 143–156, 1998.
14. O'Neill, R. V. et al. Landscape pattern metrics and ecological health. *Ecosystem Health* 5, 225–233, 1999.
15. Skole, D., and Tucker, C. Tropical deforestation and habitat fragmentation in the Amazon: Satellite data from 1978 to 1988. *Science* 260, 1905–1910, 1993.
16. Plummer, S. E. Perspectives on combining ecological process models and remotely sensed data. *Ecological Modelling* 129, 169–186, 2000.
17. Oetter, D. R. et al. Land cover mapping in an agricultural setting using multiseasonal Thematic Mapper data. *Remote Sensing of Environment* 76, 139–155, 2001.
18. Walsh, S. J. et al. A multi-scale analysis of LULC and NDVI variation in Nang Rong District, Northeast Thailand. *Agriculture, Ecosystems, and Environment* 85, 47–64, 2001.
19. Walsh, S. J. et al. Patterns of change in land use, land cover, and plant biomass: separating inter- and intra-annual signals in monsoon-driven northeast Thailand. In: Millington, A., Walsh, S. J., and Osburn, P., eds., *GIS and Remote Sensing Applications in Biogeography and Ecology.* Kluwer Academic Publishers, The Netherlands, 2001.
20. Norman, A. L. Isolating Seasonal Variation in Landuse/Landcover Change Using Multitemporal Classification of Landsat ETM Data in the Peruvian Amazon. MA thesis, University of Texas, Austin, Tex., 2005.
21. Hall, F. G., Strebel, D. E., and Sellers, P. J. Linking knowledge among spatial and temporal scales: Vegetation, atmosphere, climate and remote sensing. *Landscape Ecology* 2(1), 3–22, 1988.
22. Butzer, K., and Helgren, D. Livestock, land cover and environmental history: The Tablelands of New South Wales, Australia, 1820–1920. *Annals of the Association of American Geographers* 95, 80–111, 2005.

23. Ramos, F. A multi-level approach for 3D modelling in geographical information systems. *International Archives of Photogrammetry and Remote Sensing* 34(4), 43–47, 2002.
24. Axinn, W. G., Barber, J. S., and Ghimire, D. J. The neighborhood history calendar: A data collection method designed for dynamic multilevel modeling. *Sociological Methodology* 27, 355–392, 1997.
25. Allen, T. F. H. The landscape "level" is dead: Persuading the family to take it off the respirator. In: Peterson, D. L., and Parker, V. T., eds., *Ecological Scale: Theory and Applications*. Columbia University Press, New York, 1998.
26. Millington, A. C., Walsh, S. J., and Osborne, P. E. *GIS and Remote Sensing Applications in Biogeography and Ecology*. Kluwer Academic Publishers, Boston, 2001.
27. Crews-Meyer, K. A. Agricultural landscape change and stability in northeast Thailand: Historical patch-level analysis. *Agriculture, Ecosystems and Environment* 101, 155–169, 2004.
28. Brown, D. G., Aspinall, R. J., and Bennett, D. A. Landscape models and explanation in landscape ecology—A space for generative landscape science? *Professional Geographer*, Focus section 58(4), 2006.
29. McGarigal, K., and Marks, B. J. *FRAGSTATS: Spatial Pattern Analysis Program for Quantifying Landscape Structure*. Forest Science Department, Oregon State University, Corvallis, Oregon, 1993.
30. Jensen, J. R. *Introductory Digital Image Processing: A Remote Sensing Perspective*, 3rd ed. Prentice Hall, Upper Saddle River, N.J., 2005.
31. Congalton, R. G., and Green, K. *Assessing the Accuracy of Remotely Sensed Data: Principles and Practices*. Lewis Publications, Boca Raton, Fla., 1999.
32. Crews-Meyer, K. A. Temporal extensions of landscape ecology theory and practice: LULCC examples from the Peruvian Amazon. *Professional Geographer*, Focus section 58(4), 421–435, 2006.
33. Crews-Meyer, K. A. Integrated Landscape Characterization via Landscape Ecology and GIScience: A Policy Ecology of Northeast Thailand. Doctoral dissertation, University of North Carolina, Chapel Hill, 2000.
34. Mertens, B., and Lambin, E. F. Land-cover-change trajectories in Southern Cameroon. *Annals of the Association of American Geographers* 90, 467–494, 2000.
35. Crews-Meyer, K. A. Assessing landscape change and population-environment interactions via panel analysis. *Geocarto International* 16, 69–80, 2001.
36. Fukui, H. *Food and Population in a Northeast Thai Village*. University of Hawaii Press, Honolulu, 1993.
37. Donner, W. *The Five Faces of Thailand: An Economic Geography*. St. Martin's Press, New York, 1978.
38. Arbhabirama, A. et al. *Thailand Natural Resources Profile*. Oxford University Press, Singapore, 1988.
39. Craig, I. A., and Pisone, U. Overview of rainfed agriculture in northeast Thailand. In: *Proceedings of a Workshop on Soil, Water and Crop Management Systems for Rainfed Agriculture in Northeast Thailand*. Khon Khaen University, Khon Khaen, Thailand, 1985.
40. Crews-Meyer, K. A. Personal communication with villagers during fieldwork, January to February, unpublished, 1999.
41. Walsh, S. J. et al. Scale-dependent relationships between population and environment in Northeastern Thailand. *Photogrammetric Engineering and Remote Sensing* 65, 97–105, 1999.
42. Fox, J. et al. *People and the Environment: Approaches for Linking Household and Community Surveys to Remote Sensing and GIS*. Kluwer Academic Publishers, Boston, 2003.

43. Holling, C. S. Resilience and stability of ecological systems. *Annual Review of Ecology and Systematics* 4, 1–23, 1973.
44. Peterson, D. L., and Parker, V. T. *Ecological Scale: Theory and Applications.* Columbia University Press, New York, 1998.
45. Griffiths, R. B. Consistent histories and the interpretation of quantum mechanics. *Journal of Statistical Physics* 36, 219–272, 1984.

7 Developing a Thick Understanding of Forest Fragmentation in Landscapes of Colonization in the Amazon Basin

Andrew C. Millington and Andrew V. Bradley

CONTENTS

7.1 INTRODUCTION

Landscape fragmentation is a key concern of landscape ecologists and conservation biologists. Landscapes provide the habitats and determine the resources necessary for plant and animal species to survive. As landscapes fragment, the proportions of different elements in any landscape change, as do their spatial properties.[1] Efforts to understand spatial patterns of, and relationships between, these elements have been a key objective of landscape ecology, while conservation biologists have tried to relate the responses of species to these spatial patterns,[2,3,4] sometimes through direct manipulation of landscapes (cf. review of such experiments by Debinski and Holt[5]) but more often by observation. Many commentators on environmental issues in the humid tropics cite landscape fragmentation, along with the more general "forest loss," as major determinants of biodiversity loss and ecological deterioration.

Research to understand ecological responses to forest loss over time is fragmentary in itself, but research carried out under the auspices of the Biological Dynamics of Forest Fragments Project in northern Amazonia[6,7,8] has provided a plethora of notable exceptions, the results of which were recently reviewed by Laurence et al.[9]

Research undertaken by biologists on the ecological and conservation aspects of fragmentation of tropical forests is voluminous when compared to the number of research papers that link socioeconomic processes to the spatial phenomenon of forest fragmentation in the humid tropics. The relative paucity of research on how particular patterns of fragmentation are generated should be of concern because only when we fully understand how and why fragmentation occurs at a wide range of geographical scales will we be able to plan for, manage, and accrue conservation benefits. Research undertaken on this topic generally focuses on the spatial patterns that result from, or are the *end points* of, a generalized process (or set of processes) of tropical forest conversion.[10,11,12,13] These spatial patterns have been termed *clearance typologies* by Husson et al.[10] or *clearance morphologies* by Lambin.[11] Examples of the linkages between spatial typologies or morphologies and generalized economic activities are: planned settlement creates the so-called fishbone pattern of forest loss; spontaneous settler colonization along road networks creates linear corridors of clearance; large-scale commercial ranching and other types of commercial agriculture create large blocks of pasture, cultivation, and forest; subsistence agriculture creates a diffuse mosaic of small clearings; very high rural population densities leave an agricultural landscape with forest islands; and islands of forest surrounding urban areas occur when peri-urban plantations predominate. It is the first of these typologies—planned settlement leading to fishbone patterns of land clearance—that we address in detail in this chapter by drawing on evidence from Chapare, Bolivia.

There are limitations to these generalized process-pattern relationships. Imbernon and Branthomme[13] noted variations in the process-pattern relationships within relatively small study areas across the tropics, possibly indicating that the spatial scale at which much of this research has been carried out only allows very generalized observations of the linkages between drivers of land use change and the resulting patterns of fragmentation to be made. Lambin[11] and Hargis et al.[14] describe how spatial patterns can morph from one to another over time as fragmentation evolves, indicating that some temporal dependency might exist and that patterns may not, in themselves, be *end points*. More germane to our work is that these process-pattern generalizations typify a "thin" understanding of how particular patterns of forest fragmentation are created by the drivers (agents) of land cover change, despite the fact that the actions of such agents in the humid tropics are well understood from the syntheses that have been undertaken.[15,16,17]

A case in point is the relatively "thin" understanding between road construction and forest fragmentation that has developed for the Amazon Basin.[18] Figure 7.1 illustrates what is meant by a "thin" understanding and indicates where a "thick" understanding needs to be developed to meet the needs of conservation planning. We acknowledge that research in a limited number of colonization zones in the Amazon Basin has modeled land colonization.[19,20,21,22] The foci of these studies has generally been on societal processes and impacts and on the resulting forest cover in a *general* sense. Although this has deepened our understanding of land use dynamics

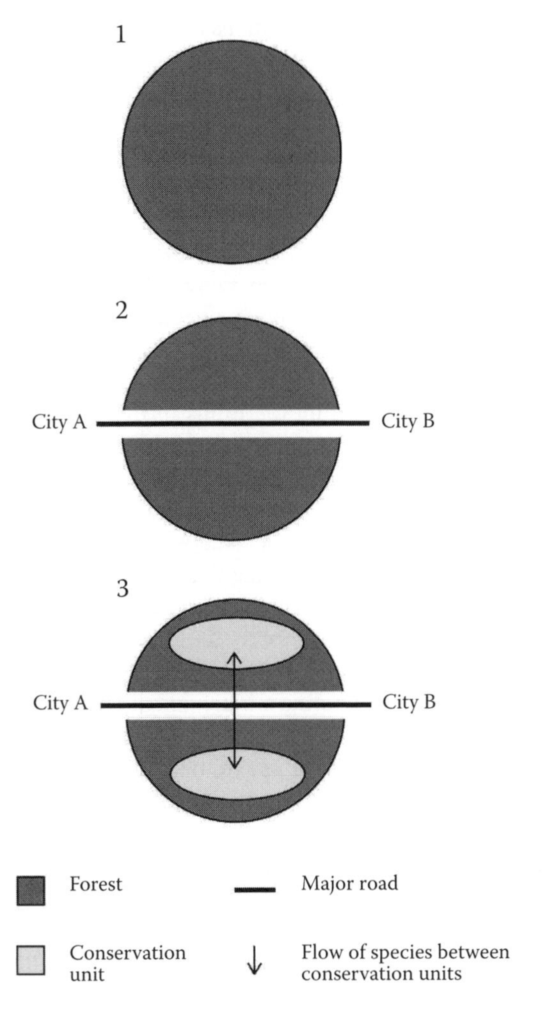

FIGURE 7.1 Thick and thin understandings of forest fragmentation along roads in lowland forests of the Amazon Basin. The progression from stage 1 to 2 shows a forest block dissected by a road connecting cities A and B. Stage 3 illustrates large-scale fragmentation between two conservation units and the connectivity between them that is required is illustrated by the black arrow. A "thin" understanding of fragmentation is represented in Stages 2 and 3; the development of the "thick" understanding that we argue for in this chapter is required for the white area along the main road.

in colonization zones, knowledge of the causal linkages between processes that lead individual land owners to fragment the forests on their properties in particular ways, and how the fragmentation patterns on individual properties mesh together across a community or a number of communities, has rarely been investigated, though its importance is recognized.[23,24] A notable exception is the research by Perz and Walker[25] who applied a neo-Chayanovian analysis to secondary forest regrowth on small colonist farms, arguing that more attention needs to be paid to households as the most proximate context in land use decision making.

If planning and management interventions are to be made during forest conversion in areas undergoing colonization, then a detailed understanding of how agents are operating in the landscape at different geographical scales is essential. For example, the influence of roads and other lines of access occur at one scale. Generalizations about the environmental impacts of roads at this scale have been recognized[26,27,28] and used, somewhat contentiously, to model the impact of development policies in Brazil.[29,30,31,32,33] But nested below the road network in geographical space is almost always a cadastre or land property grid. Although the roads can be constructed both before and after a cadastre has been surveyed, we argue that the roads *and* the cadastre provide two spatial imprints connected through a scale hierarchy in forested landscapes that are destined to fragment. Moreover, attempts to model fragmentation spatially based on road building have missed a fundamental point. That is, it is the colonist households within the limits of their properties that create the patterns of forest fragmentation by responding to economic and policy signals with machetes and chain saws, rather than planners and road builders with maps and bulldozers. It could be argued that the planners and road builders spatially constrain what farmers can clear, as well as provide the wherewithal to extract timber and produce from their farms. We acknowledge that the argument that we make here may only apply to farmers in planned colonization schemes: it may not apply to other types of humid tropical forest colonization in the Amazon Basin or elsewhere.

Developing a "thick" understanding of fragmentation at contemporary deforestation fronts therefore requires integrating the actions of land managers on individual properties over time and meshing them together within the road networks and land property grids. In an applied vein, what is required specifically to plan and manage landscapes of colonization is research into the spatial and temporal dynamics of forest fragmentation, which cover multiple scales, considering all agents of change, and the links between agents and scales. The results of such research will allow an important question to be answered. That is, how do the collective actions of land managers in a particular area lead to particular patterns of forest fragmentation over trajectories of time? If this question can be answered robustly, then two further questions of concern to landscape ecologists and conservation planners can also be tackled:

1. Can zones of colonization be planned so that they can develop into multipurpose landscapes that allow rural production systems to co-exist with biological conservation?
2. How can existing, partially fragmented landscapes be planned for?

In this chapter we explain how we have attempted to develop a "thick" understanding of the dynamics of landscape fragmentation in a colonization zone in the lowland humid tropics of Bolivia, and then reflect on further research needs.

7.2 CASE STUDY AND METHODS

7.2.1 CHAPARE

We used observations from the Chapare region of Bolivia in this research. Chapare is a colonization zone in the humid tropical lowlands of Bolivia dating back to the

TABLE 7.1
Salient Information Concerning the Three Communities Studied

Community	Area (ha)	Altitude (m.a.s.l.)	Number of properties	Year of first settlement	Prevailing economic activities 2000–2003
Arequipa	1,220	250	60	1983	Banana, black pepper, cassava, heart of palm, and rice cultivation
Bogotá	3,196	250–350	90	1972	Cattle rearing: beef and dairy
Caracas	1,745	220	110	1963	Banana, mandarin, and orange cultivation

1930s, though most colonization and forest conversion has taken place since the 1960s.[34,35,36,37,38] The area is bounded to the north by relatively undisturbed lowland tropical forests and to the south by montane forests. These forests are likely to remain relatively undisturbed in the foreseeable future because to the north they are either permanently or seasonally inundated, and to the south they are protected by Parque Nacional (PN) Carrasco. The zone of colonization creates a wedge of livelihoods and disturbance between these two forest blocks, thereby compromising the exchange of animals and, less obviously, plant material between the two. Given the strong affinities between the animals and plants in the lowland montane forests in PN Carrasco and the lowland forests, this is a cause of concern for conservationists.

Chapare benefits from a dense network of primary and secondary roads augmented by foot tracks.[34,37,39] This network has developed progressively since the 1960s in two ways. First, by its physical extension; second, by upgrading the road surfaces from dirt to tarmac or cobble. A land property grid has developed in parallel with the road network, and the two are integral to colonization of the area. The land now occupied by each of the communities in Chapare was surveyed and marked out down to the limits of each land parcel by the Instituto Nacional de Colonización (INC) before it was settled. Titles were given to colonists moving into each community. These records are held by INC. The transportation network and the land property grid combine to provide the spatial stage on which colonists act out their livelihoods, while simultaneously spatially constraining their activities.

7.2.2 Methods

To understand the spatial and temporal relationships between the different agents of change, one of us (Bradley) conducted detailed surveys in three communities between 2000 and 2003.[34] Salient details of each community are listed in Table 7.1. In each community the following research was undertaken to develop an understanding of the spatial and temporal dynamics of land cover change:

1. The land property grid for each community was obtained from INC and the owners of each property identified.
2. Permission was obtained from the community *sindicato* to interview land owners/managers. Subsequently those interviewed were selected randomly.

3. Land cover maps of each community were created using Landsat MSS, TM, and ETM+ imagery acquired in the dry seasons of 1975, 1976, 1986, 1992, 1993, 1996, and 2000 using classification algorithms in ERDAS Imagine (full details of which can be found in Bradley[34]). Field verification was carried out on the maps derived from imagery acquired in 2000 (Figure 7.2). These maps were then simplified into binary forest and nonforest covers.
4. Each land owner/manager selected was shown the time sequence of forest/nonforest maps for their property and asked to recall aspects of forest clearance and what crops had been grown at the times the images were acquired. This was done using participatory rural appraisal methods, the most informative of which was to walk each farmer's property with him. This enabled farmers to verify their recall of what had been grown at particular times, and also enabled geolocation of these observations using a global positioning system (GPS) receiver in nondifferential mode.
5. For each property surveyed, a forest/nonforest map—*a property forest/nonforest map*—was annotated with the owner/manager's observations.

Properties are typically 20 ha in areas of cultivation and 50 ha in areas of livestock rearing. Fifteen, 13, and 17 properties were surveyed in detail for Arequipa, Bogotá, and Caracas, respectively (Table 7.1). The names of the communities and the farmers we interviewed have been made anonymous in accordance with normal social science survey practices, and because some of the farmers have illegally grown coca in the past. The observations made about farms and farmer's responses to questions were used in two ways. First, to understand the drivers of land use change in Chapare from the 1970s to the present time,[34] and second, to verify the forest/nonforest maps of each community—*community forest/nonforest maps*—that the *property forest/nonforest maps* of each property surveyed were extracted from. The verified *community forest/nonforest maps* were then used to map areas of forest and nonforest for each community.

7.3 A CONCEPTUAL MODEL OF FRAGMENTATION

By comparing the progressive development of spatial patterns of forest fragmentation between the three communities we developed a conceptual model of forest fragmentation that has six phases (Figure 7.3). The first or planning phase occurs before colonization, and, consequently, no forest has been cleared at this time (Figure 7.3a). However, this stage is important for it is at this time that the general spatial configuration of forest and agriculture patches that will ultimately populate this geographical space is determined. This is because the property grid is surveyed and laid out, and the (unimproved) access roads and tracks are constructed. This phase exists for a short period of time before colonists arrive. In the second phase—early colonization—olonists clear forest at the primary ends of each property, thereby creating a simple pattern of fragmentation on either side of the primary access road (Figure 7.3b). In the three communities we researched intensively, all properties were occupied almost immediately, and the areas of each plot cleared (calculated from *community forest/nonforest* maps) were similar because the colonists either arrived together or within a few months of each other. Moreover, their motivations

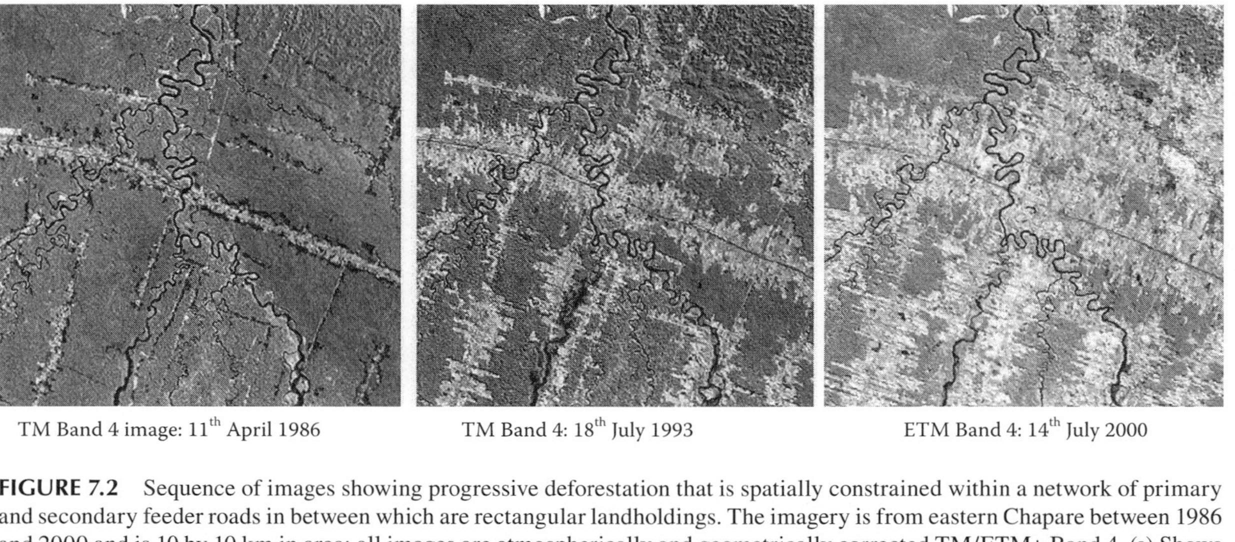

TM Band 4 image: 11th April 1986 — TM Band 4: 18th July 1993 — ETM Band 4: 14th July 2000

FIGURE 7.2 Sequence of images showing progressive deforestation that is spatially constrained within a network of primary and secondary feeder roads in between which are rectangular landholdings. The imagery is from eastern Chapare between 1986 and 2000 and is 10 by 10 km in area; all images are atmospherically and geometrically corrected TM/ETM+ Band 4. (a) Shows the least forest clearance and was acquired on 11 April 1986. The dark gray tones that dominate the image are different types of lowland tropical forest. The main Cochabamba-Santa Cruz bisects the image and forest is cleared about 500 m on either side of the road. A network of primary feeder roads can just be seen in the forested areas on either side of the main road, although there is very little clearance in any of the communities these roads serve. The black sinuous pixels are small rivers that bisect Chapare and emanate from the Andes foothills to the south of the images. (b) Image was acquired on 18 July 1993. The light gray areas along the primary feeder roads are areas of clearance. (c) Image was acquired on 14 July 2000 and shows further deforestation. Different size forest patches that have been isolated by deforestation fronts coalescing from different directions can be seen in the center of the image. The castellated nature of deforestation in this type of colonization scheme can be clearly seen.

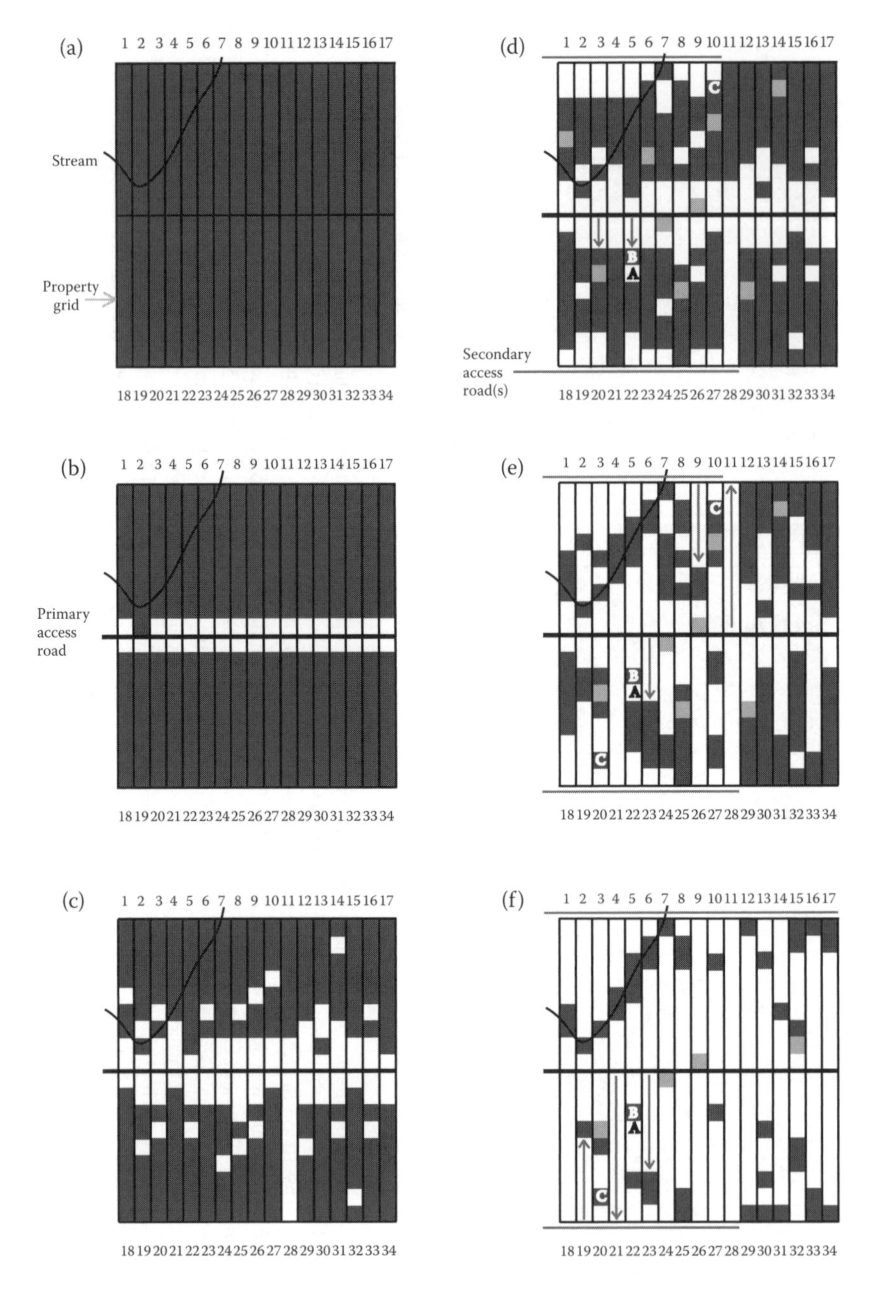

FIGURE 7.3 Six-phase conceptual model of forest fragmentation based on a community with 34 plots (numbered 1 to 34) of equal area. A primary access road (black) runs through the center of the community and a stream cuts through properties 1 to 7. Images represent: (a) the planning phase, (b) the early colonization phase, (c) the illicit coca phase, (d) the improved access phase with secondary roads (gray roads), (e) the complex clearance phase, and (f) the plot exhaustion phase. The light gray cells in phases (e) and (f) represent secondary regrowth forest.

for clearance were similar, that is, to clear forest for land to plant subsistence crops (followed by cash crops in subsequent years) and to acquire construction materials for houses. In a less-detailed examination of the image series for all of Chapare, we saw this phase replicated in all communities.

More complex spatial patterns of forest fragmentation establish themselves in the third phase of the model—the illicit coca cultivation phase (Figure 7.3c). Because coca cultivation in the lowlands of Bolivia has always been illegal (in comparison to cultivation for chewing in the subtropical montane forests where it is legal), farmers generally adopted strategies to cultivate coca that fragmented forests in particular ways. However, in the 1970s when coca cultivation and cocaine production was barely controlled by the government, coca was grown openly at the primary ends of many properties. As government crackdowns on coca growing took effect, many farmers grew *legal* cash and subsistence crops at the primary ends of their plots and retreated into the remaining forest on their properties to clear small areas to grow coca. This was the main cause of forest perforation, and the extent of perforation depicted in this model (Figure 7.3c) is high because of illegal coca cultivation and is probably greater than it would be in other colonization areas. This assumption has yet to be tested. The colonist footprint model developed by Brondizio et al.[20] predicts high rates of forest clearance at this stage as farmers prepare land to plant perennial cash crops. But our evidence indicates that although a few farmers cleared forest at much faster rates than others (e.g., property 28, Figure 7.3c), this was exceptional because the vast majority of farmers only had to clear small areas to cultivate the perennial crop of choice—coca—which, because it has a high-selling return, is conservative in its land requirements. Differences in forest clearance rates between individual properties occur at this stage because few farmers cultivated land-hungry perennial crops at this time, as predicted by Brondizio et al.,[20] rather than coca. These differences lead to the castellated pattern of the forest/nonforest boundaries that characterize "fishbone" deforestation.

The development of secondary, unimproved feeder roads at some point in time during colonization is typical of most communities in Chapare. Roads and tracks are constructed along the boundaries of communities to connect with the roads that were constructed initially. We have characterized this phase as improved access, and in Figure 7.2d secondary feeder roads are drawn along the secondary end of properties 1 to 11 and 18 to 27. The establishment of such roads and tracks allows plots to be cultivated from both ends, but the actual reasons for their construction are unclear at present. Our interviews so far suggest they may be constructed to consolidate community boundaries, but they may simply improve access. Whatever the reason, they can be used to split up properties to satisfy actual and potential disputes over inheritance, or allow farmers to cultivate more fertile soils at one end of their property while allowing recovery of vegetation and soil at the other end of the plot.

In the fifth phase of the model—complex clearance—a significant amount of the land in a community is under some type of cultivation (Figure 7.3e). The term complex arises because many landscape ecology metrics attain their highest values during this phase. The formation of both forest patches and the extension of the forest perimeter are due to differential rates of clearance between farmers, and the continuation of forest clearance from both the primary and secondary ends of some properties. The formation of a forest patch due to the differences in rates of forest clearance is shown in Figures 7.3d and 7.3e. The farmers in properties 21 and 23 have cleared forest at faster rates than the farmer in property 22. As a consequence a patch of forest (B) on property 22, which once shielded a coca clearing (A), is now surrounded by agricultural land. This method of patch formation is commonplace in Chapare, and occurs because of the intersection of government coca eradication policies (which causes perforation-style clearance of forest deep in properties), differences in crop choices between farmers (which leads to different land requirements to grow particular crops), and differences in household circumstances and aspirations. The creation of forest patch C on property 10 is a variant on the way in which forest patch B was formed. In this case the rates of forest clearance between properties 9, 10, and 11 are not only different in the amounts of forest cleared annually, but also the directions of clearance are different because of the influence of the secondary access road on property 9.

We have termed the final phase plot exhaustion (Figure 7.3f). By this we mean that most of the forest has been cleared. Some isolated patches remain, and there are also patches of secondary regrowth forest and forests in areas that are difficult to clear or are located on land that cannot be cultivated (e.g., the riparian forest in properties 1 to 7).

We have evidence that properties are already changing hands by the time the penultimate and final stages of the model are reached. Although we have not recorded land being sold in the 45 properties we have surveyed in detail, we have come across this on farms we have visited in other communities. Some properties are being sold to new owners and some wealthy farmers purchase adjacent plots to increase their contiguous land holdings. We have indicated this in the model by combining properties 28 and 29 in Figure 7.3f.

7.4 BEHAVIOR OF LANDSCAPE METRICS

The conceptual model outlined above is based on detailed observations made in three communities, and to evaluate its utility for analyzing the ecological implications of fragmentation we used metrics commonly employed by landscape ecologists and conservation biologists. We calculated proportional forest cover, the number of forest patches, and the forest/nonforest edge length for each phase in the model (Table 7.2). These data are visualized in Figure 7.4.

Lambin[11] postulated that landscape metrics used to characterize fragmentation would follow a particular trajectory as tropical forest landscapes changed from those that were entirely forested, through landscapes of agricultural patches in a forest matrix, to entirely agricultural landscapes (i.e., a few forest patches in an agricultural matrix). He did not quantify this postulated behavior, but hypothesized that the metrics would attain peak values in the heterogeneous, intermediate landscapes and would be low for homogenous forest or agricultural landscapes. Trani and Giles[40] simulated deforestation of a hypothetical forest and calculated metrics at various points along a deforestation/fragmentation trajectory. Three metrics from their analysis—mean forest patch size, the forest/nonforest edge length, and the mean nearest neighbor distance between forest patches—are shown in Figure 7.5. We calculated the same landscape ecology metrics as Trani and Giles[40] for each community we studied using Fragstats.[41] As the metrics followed similar trends in each community, we only illustrate the metrics for Communidad Arequipa in this

TABLE 7.2
Selected Landscape Metrics Calculated for the Six Phases of the Conceptual Model

Phase in conceptual model	Proportional forest loss (%)	Number of forest patches	Number of cultivation patches	Mean patch size (nominal units)	Forest edge length (nominal units)
Planning	0	1	0	306	0
Early colonization	10.8	2	1	137	37
Illicit coca	29.7	4	20	71	140
Improved access	36.6	6	15	49	151
Complex clearance	58.8	12	7	11	203
Plot exhaustion	83.3	20	2	3	135

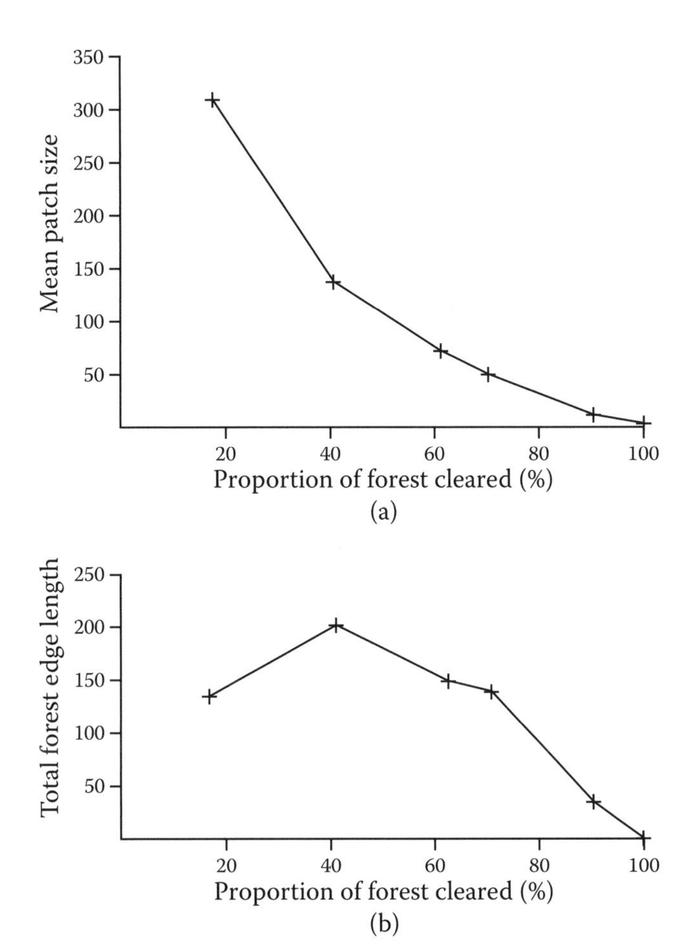

FIGURE 7.4 Metrics calculated for the conceptual model: (a) mean forest patch size, and (b) total edge length. Both are in nominal units.

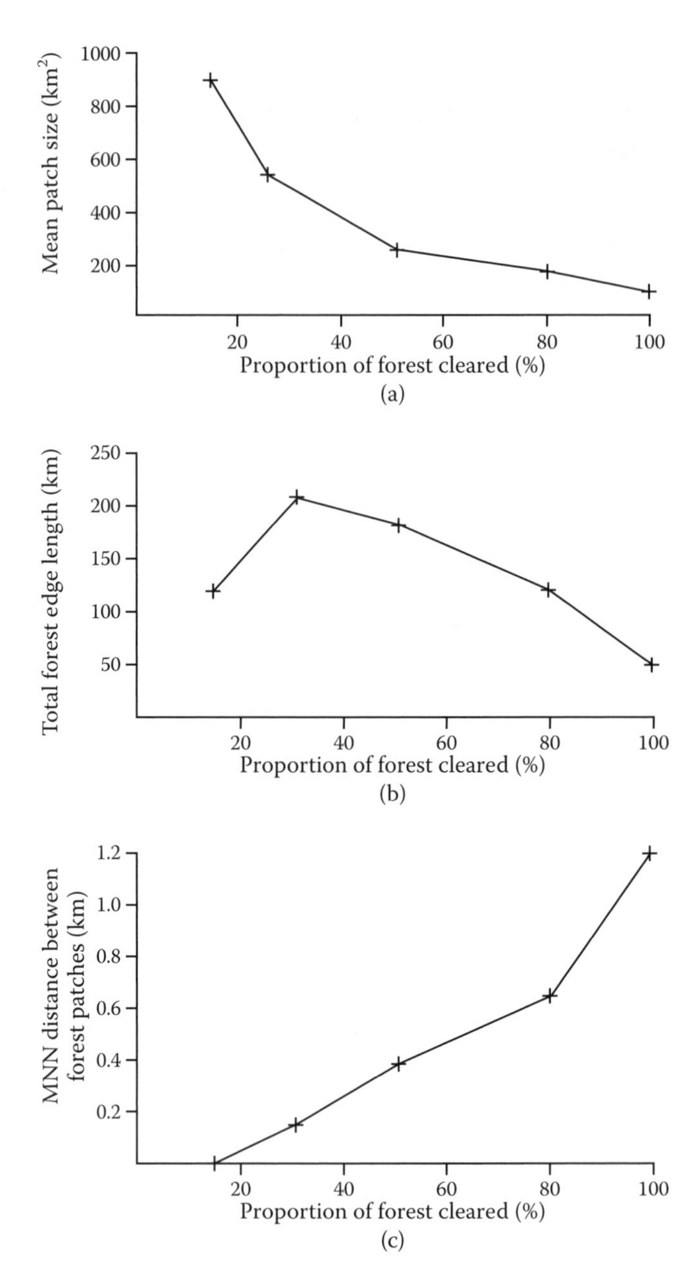

FIGURE 7.5 Selected metrics from Trani and Giles[40]: (a) mean forest patch size in km^2, (b) total edge length in km, and (c) mean nearest neighbor distance between forest patches in km.

chapter (Figure 7.6) and, as forest loss had only reached 46% by 2000 for this community, the data do not extend to very high proportional forest losses. Comparing the limited number of metrics in Figures 7.4, 7.5, and 7.6 provides an initial test of the robustness of the conceptual model. A decline in forest cover over time in the conceptual model is clear in Table 7.2. This allows the successive phases of the model to be parameterized as proportional forest losses in the graphs in Figure 7.4. Both Lambin[11] and Trani and Giles[40] used forest cover in their graphs.

Mean patch size declines in a consistent manner (Figures 7.4a, 7.5a, and 7.6a). Initially the decline is rapid, more so in reality (in Communidad Arequipa) than in the model or the simulation. At around half the area deforested, the rate of decrease in mean patch size declines and then the rate of decrease in forest patch size tapers off. Total edge length initially increases as forests begin to be cleared, and the edge length is at its greatest at intermediate forest covers and then declines. This is evident in the conceptual model, the simulation, and in the real data, Figures 7.4b, 7.5b, and 7.6b, respectively. There are differences in the peak values of total edge length, which suggest the variation in the peak may be related to the spatial configuration of fragmentation. Whereas Trani and Giles[40] simulated a somewhat random pattern of fragmentation, that in Communidad Arequipa and the simulation model are for regularly structured landscapes, which, because properties are long and thin, has a tendency to have high edge lengths at intermediate forest losses compared to lower edge lengths in more randomly fragmented forests. The third metric—mean nearest neighbor (MNN) distance between forest patches—has only been calculated for the actual data from Communidad Arequipa and extracted from Trani and Giles's[40] simulation. For this metric there is a difference in behavior. There is relatively little variation in the MNN distance in Communidad Arequipa, and the distances over the range of forest covers in the community describe a shallow U-shape. However, Trani and Giles's simulation describes an upturned-U distribution in MNN distance as forest is progressively lost.

In summary, our analysis of landscape metrics suggests that:

1. Mean patch size (MPS) and total edge length (TEL) are very robust and consistent measures of fragmentation. Although in all three cases there are similar trends in MPS and TEL with progressive deforestation and increasing forest fragmentation, there are differences in the precise nature of the curves, which may be a function of the effect that the spatial configuration of land properties have on fragmentation. From the viewpoint of model validation the latter point is not that important, but it does lend weight to the argument that understanding land owner's decisions at the small scale is important for developing our understanding of fragmentation and future planning of colonization zones.
2. The behavior of the mean nearest neighbor distance between forest patches does, however, vary between the studies and is due either to the differences in spatial scale between the two studies or the nature of forest loss. In Communidad Arequipa the spatial imprint of the cadastral grid and the ways in which people clear forest within the grid may lead to a restricted set of possible distances between forest patches and, as these have only been observed in the early stages of fragmentation, they may change significantly as more forest is cleared and fragmentation proceeds.

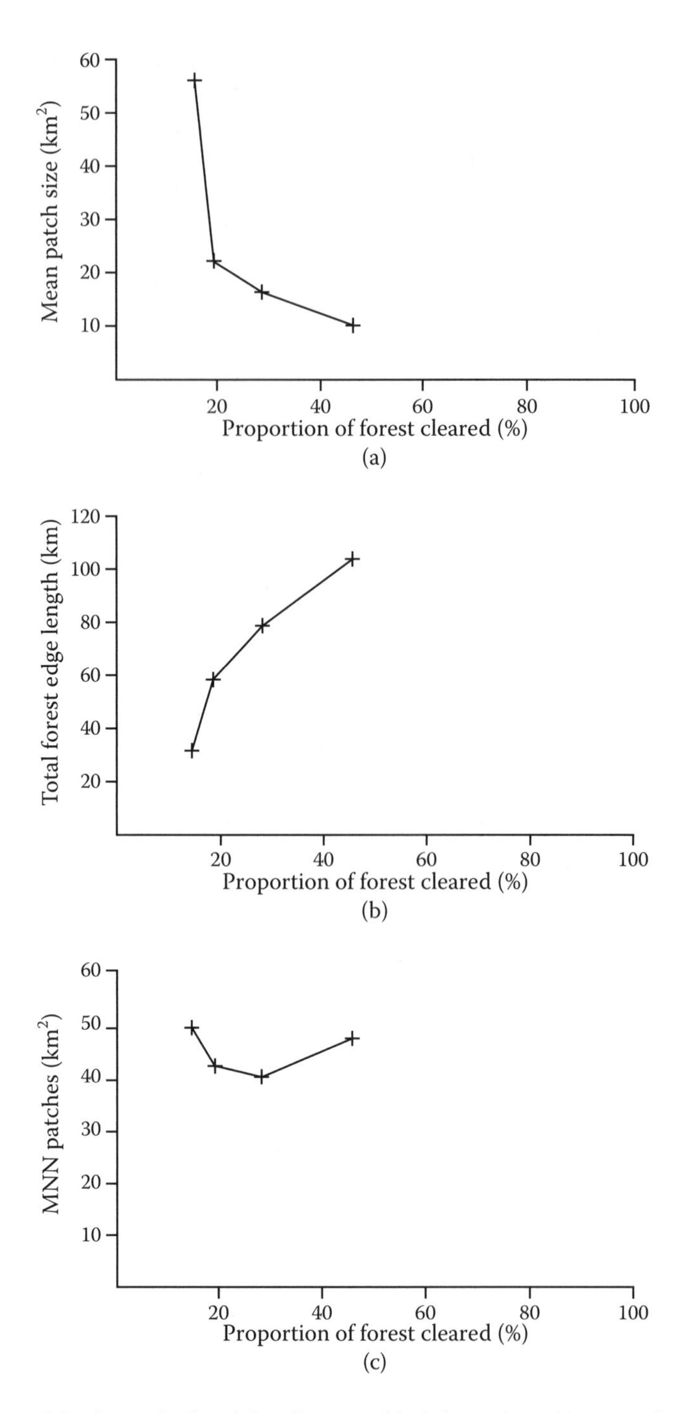

FIGURE 7.6 Metrics calculated for Communidad Arequipa: (a) mean forest patch size in km², (b) total edge length in km, and (c) mean nearest neighbor distance between forest patches in km.

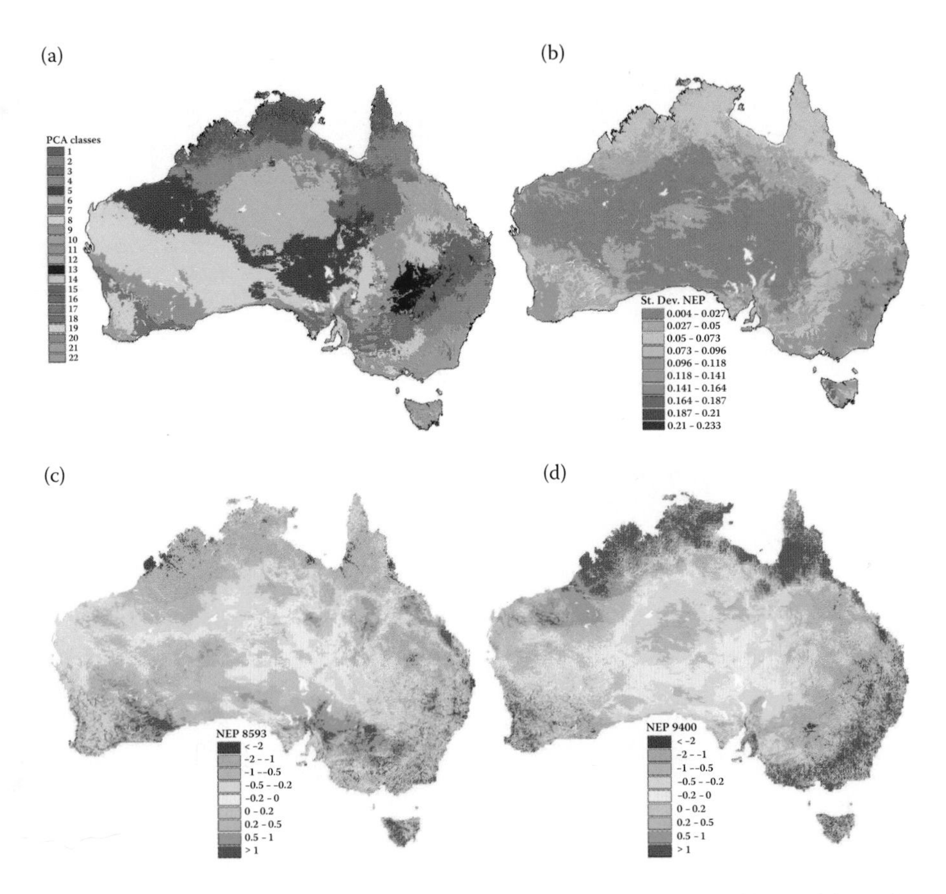

FIGURE 2.4 (a) The spatial pattern of temporal signals may be grouped by applying principal components analysis to the time series and creating a classification based on the major principal components. (b) The temporal metrics may be calculated on the spatial times series to create maps of, for example, standard deviation of net ecosystem productivity (NEP). (c) A running integration, trend, or wavelet analysis may define periods of distinct behavior in the time series that can then be summarized by metrics such as an integral of NEP for periods of decline and increase. Shown for 1985–1993 and 1994–2000 here.

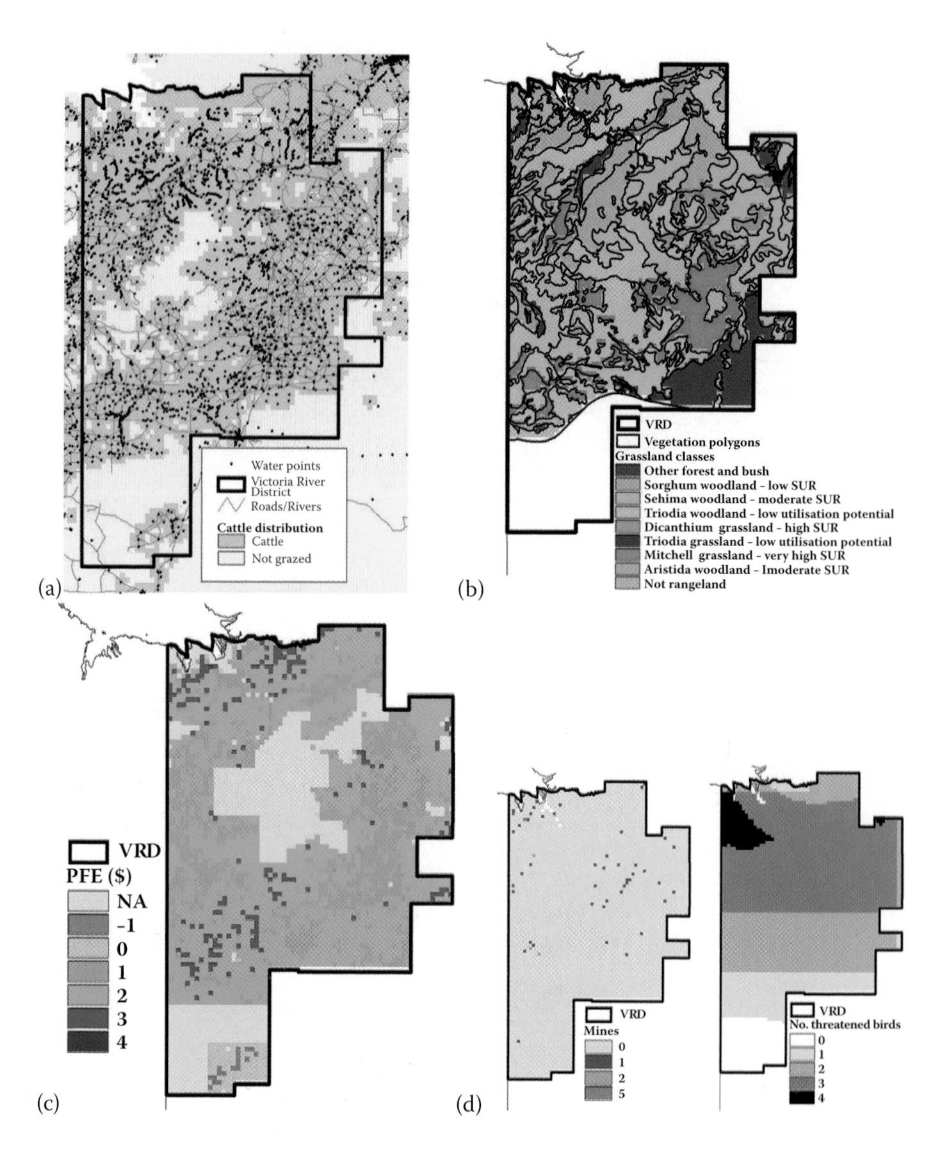

FIGURE 2.5 Temporal signals are usually based on biophysical or human phenomena that operate at a large scale (e.g., climate, interest rates). Demographic changes at a fine scale may have scale limitations due to level of aggregation in reporting. Temporal signals and indicators are filtered by spatial variation. The Victoria River District in the Northern Territory of Australia is highly productive. (a) Cattle are distributed of freehold-leasehold land but confined by water points. (b) Both productivity and ecological impact vary with vegetation type, which is associated with soils, topography, and rainfall gradient. (c) Costs are low and enterprises are profitable but the increment is small on a per hectare basis. (d) Mining with major physical disturbance occurs sporadically across the area. There are threatened bird species in the region and these may be ground nesting and impacted by grazing.

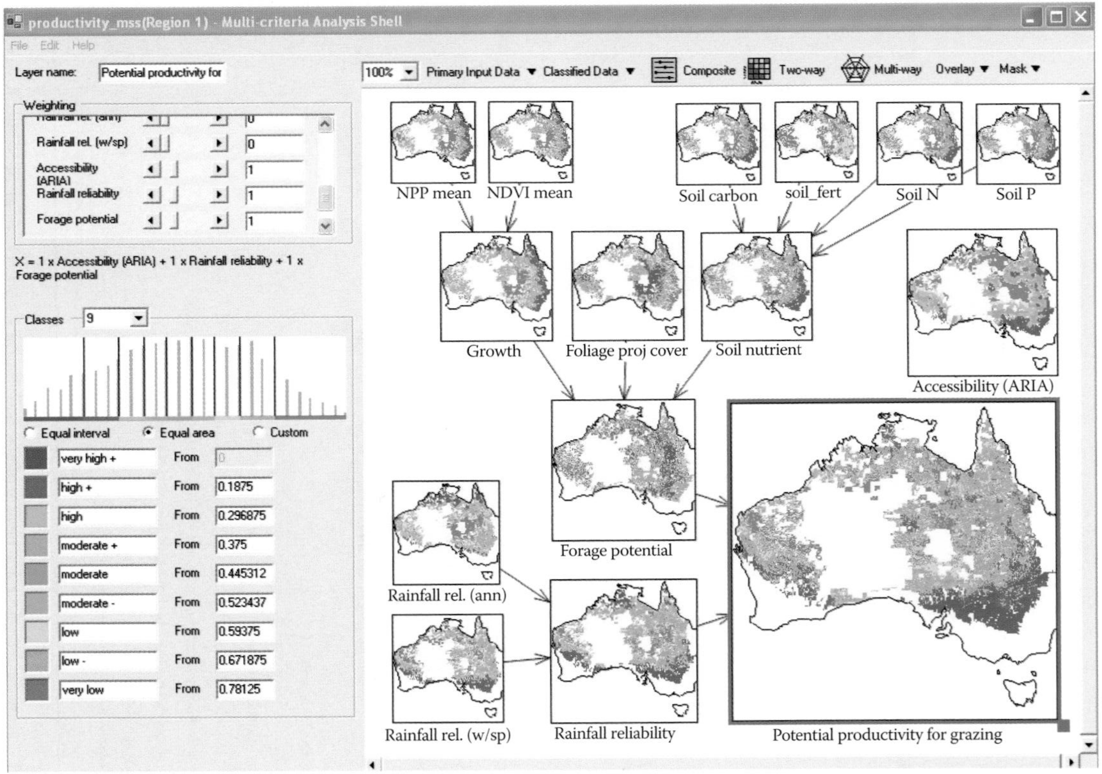

FIGURE 2.7 Cognitive mapping interface suitable for combination of diverse spatiotemporal metrics, indices, and data layers describing meaningful properties of a system under analysis. This example shows the construction of a composite index to represent potential grazing productivity from rangelands.[46,58]

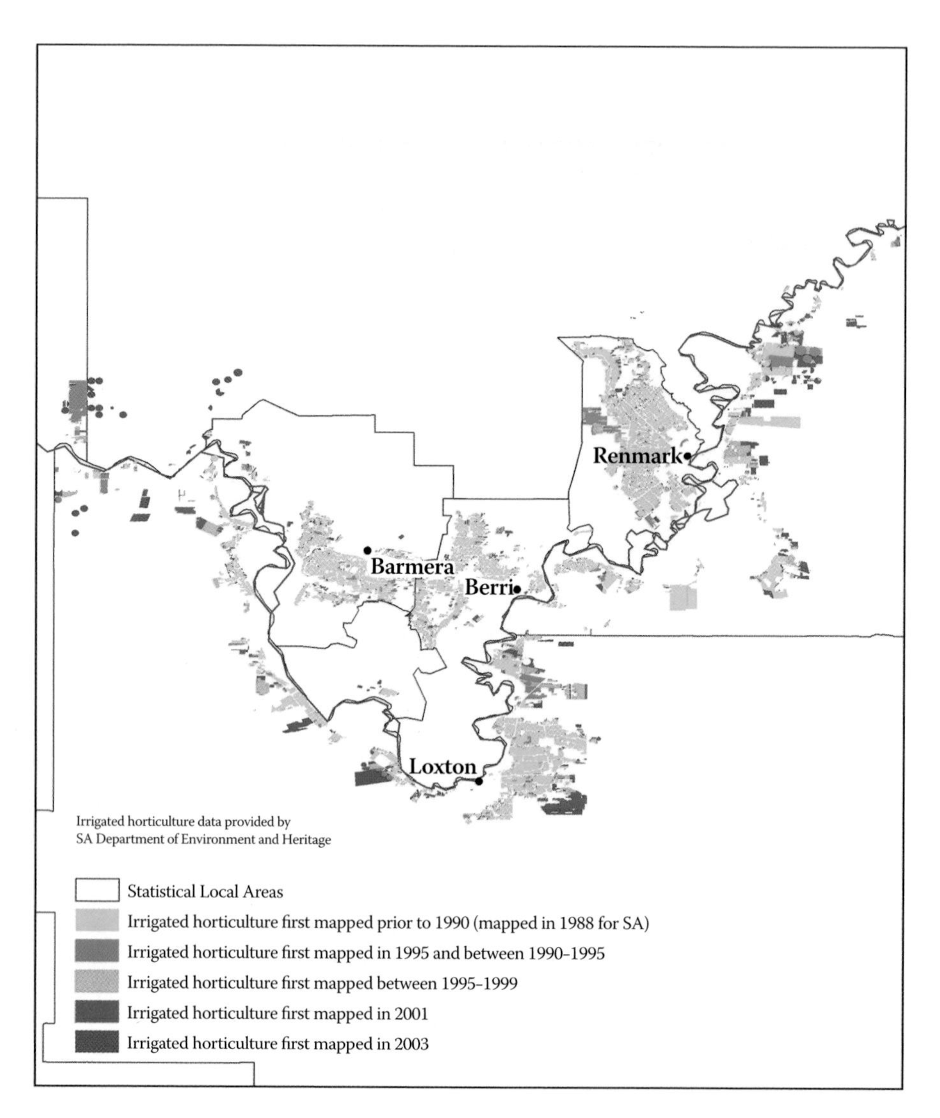

FIGURE 3.1 Land use change in the Barmera, Berri, and Renmark areas of South Australia.

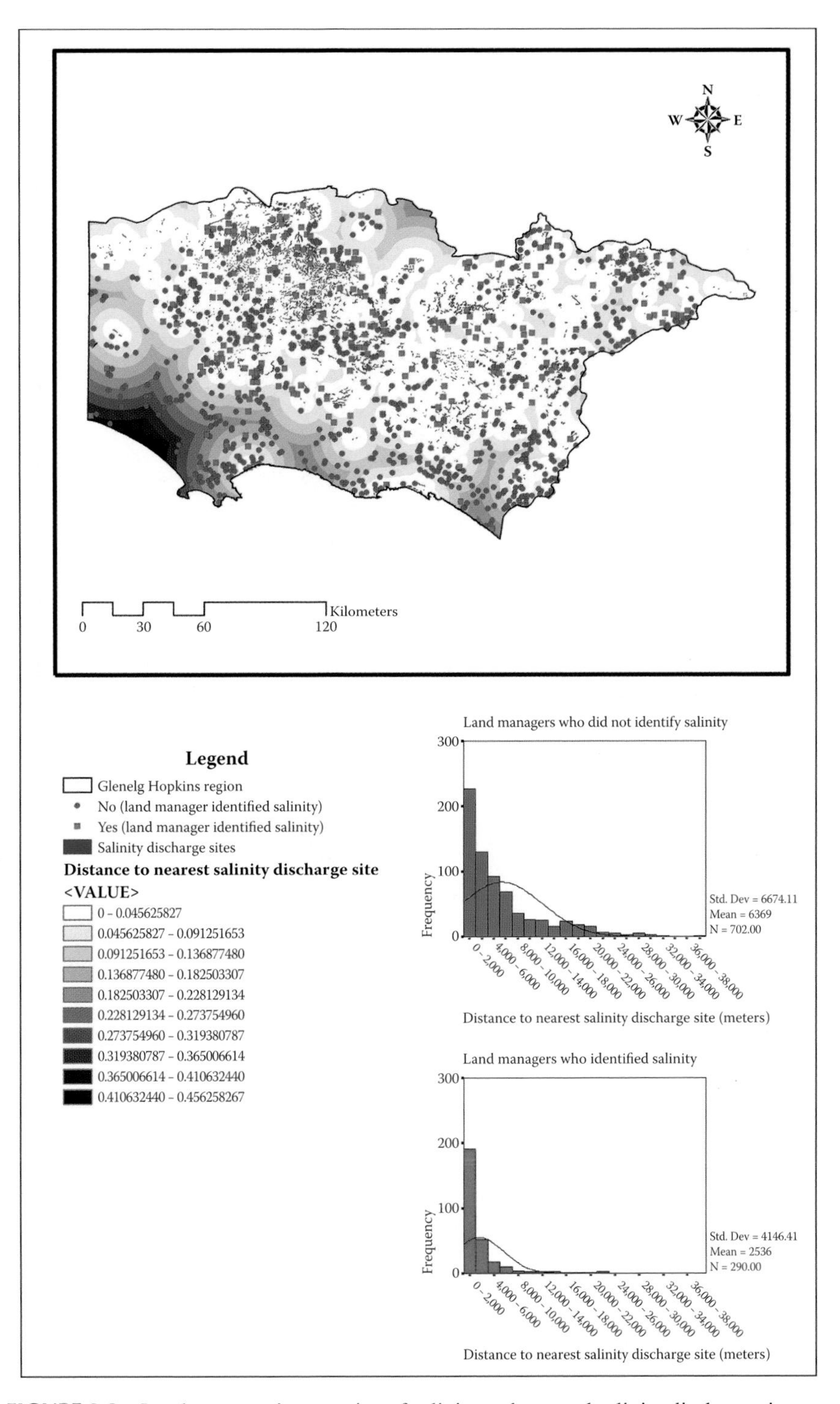

FIGURE 3.3 Land managers' perception of salinity and mapped salinity discharge sites.

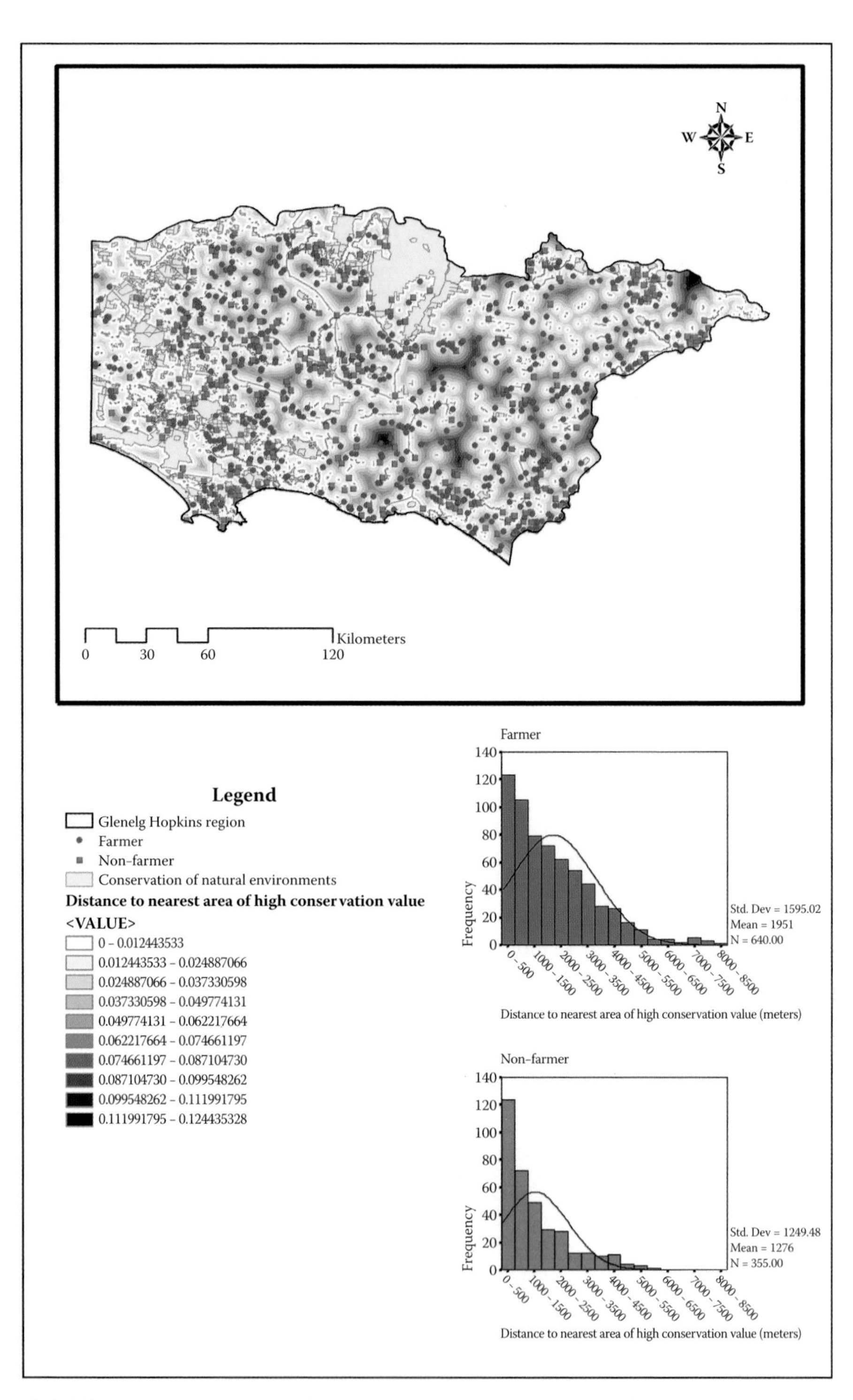

FIGURE 3.4 Land managers who manage properties near areas of high conservation value.

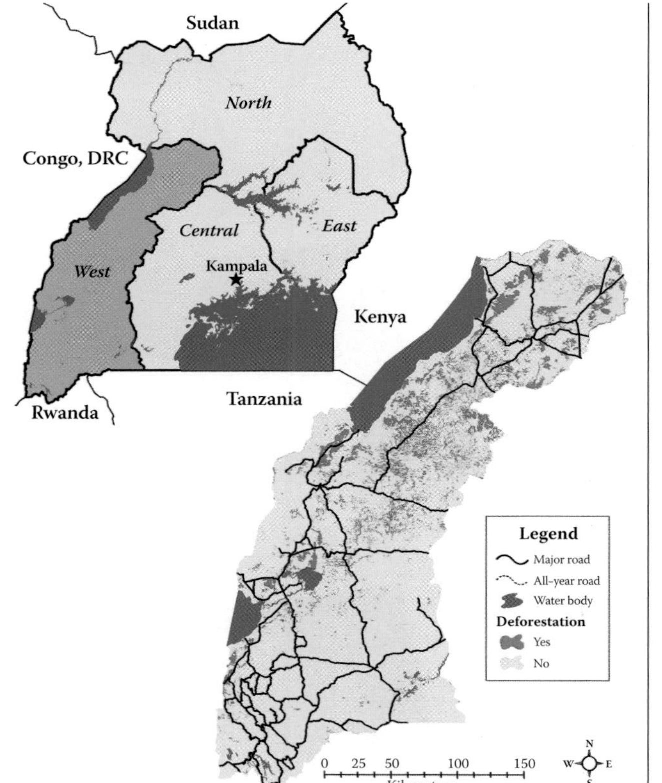

FIGURE 4.1 Uganda study area showing the distribution of deforestation within the western region of the country.

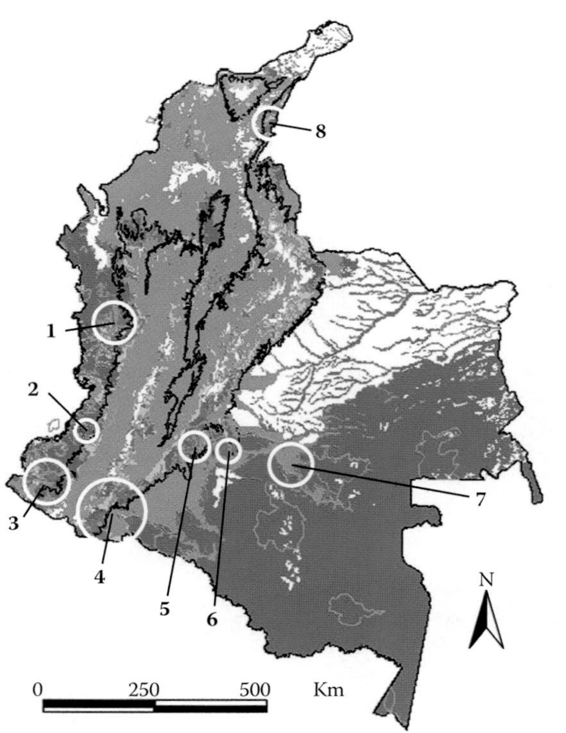

FIGURE 5.3 Predicted deforestation hot spots obtained by combining areas predicted to have the highest probability of forest conversion (>70%) from the best model (the region-specific classification tree) with the areas with greater than 2% rural population growth rate (1985–1993). Red depicts the deforestation hot spots (areas with >70% probability of forest conversion and >2% rural population growth). Orange and red depict areas with >70% probability of forest conversion. Green depicts forested areas, gray represents cleared forested areas, and white represents nonforested areas. White circled areas indicate current hot spots of deforestation, which are also areas of high-value biodiversity value: (1) Quibdó-Tribugá, (2) Farallones-Micay, (3) Patía-Mira, (4) Fragua-Patascoy, (5) Alto Duda-Guayabero, (6) Macarena, (7) Guaviare, and (8) Perijá. Black line is the Andean region, and light green lines are national parks. (From Etter et al.[50] With permission.)

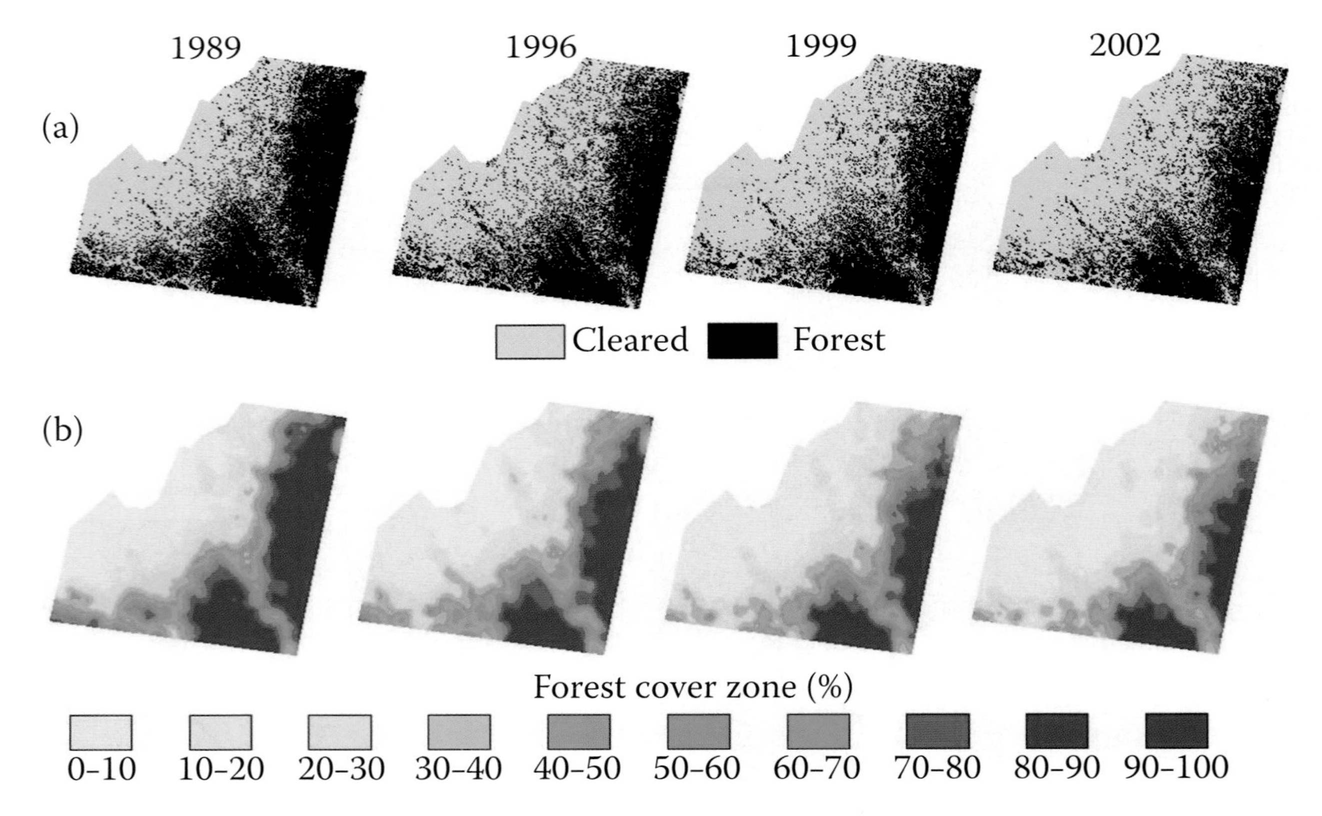

FIGURE 5.4 Forest maps of the colonization front for each study date: (a) extent of forest cover (black = forest); and (b) percentage forest cover at 10% increment zones. (From Etter et al.[39] With permission.)

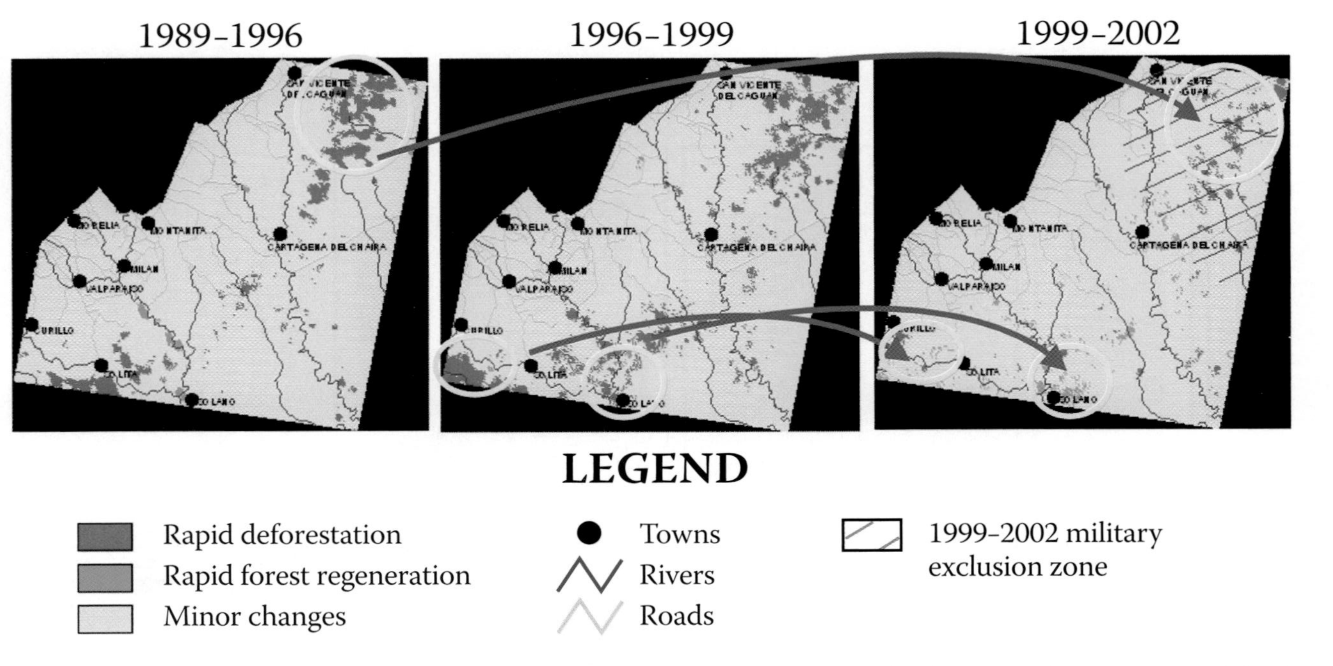

FIGURE 5.5 Spatial location of the local hot spots of deforestation (red) and regeneration (green) for the three time periods of study. (From Etter et al.[39] With permission.)

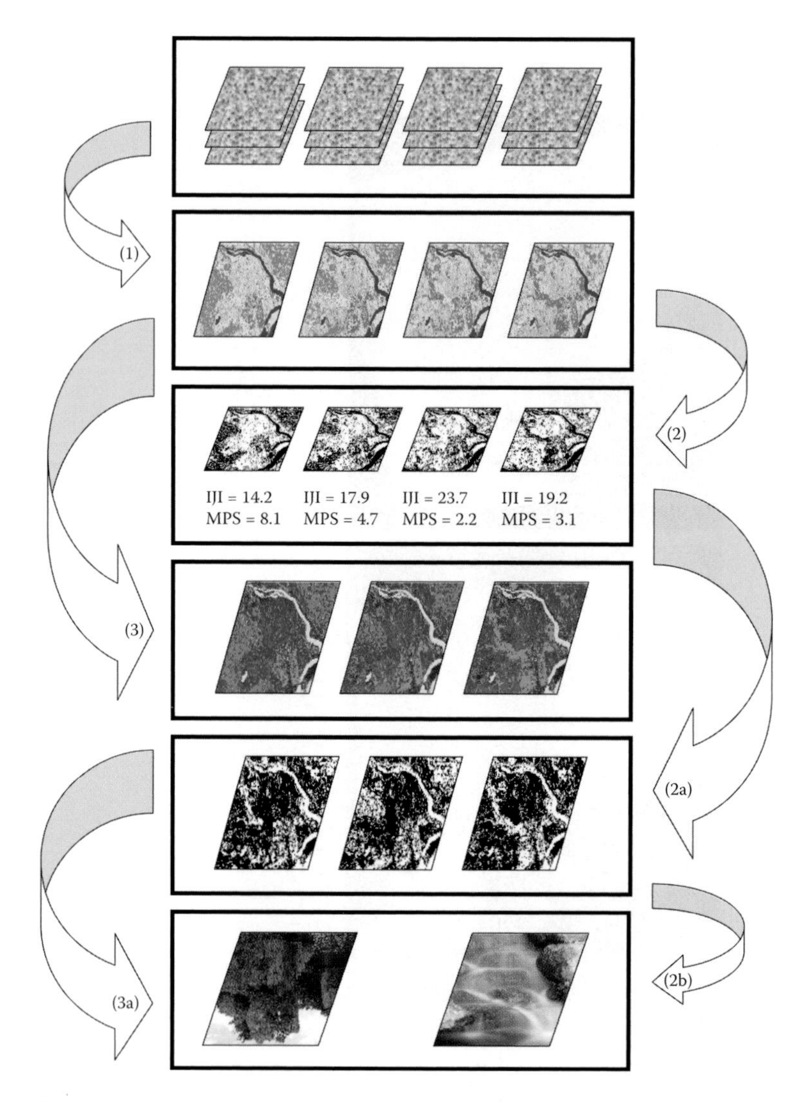

FIGURE 6.1 The panel process, conducted at both the pixel and patch levels: (1) four multispectral satellite images are each categorized into a thematic LULC classification; (2) pattern metrics are run on each of the four LULC classifications, each producing a set of patch, class, and landscape statistics (here the interspersion/juxtaposition index [IJI] and mean patch size [MPS] are shown) as well as an output image of the delineated patches; (2a) pattern metric output for each of the four times is used to calculate three piecemeal change maps for each pattern metric and each consecutive pair of images (e.g., showing fluctuations in IJI or MPS between two time periods) as per Crews-Meyer[11,27]; (2b) three pattern change maps are stacked into one panel of all structural change for each given metric (e.g., showing fluctuation in IJI or MPS through all time periods) as per Crews-Meyer[11,27]; (3) three thematic change maps are created for each of the time periods represented by the four classifications; (3a) the three thematic change maps are stacked to represent the full record of all thematic change across the four classifications as per Crews-Meyer.[3]

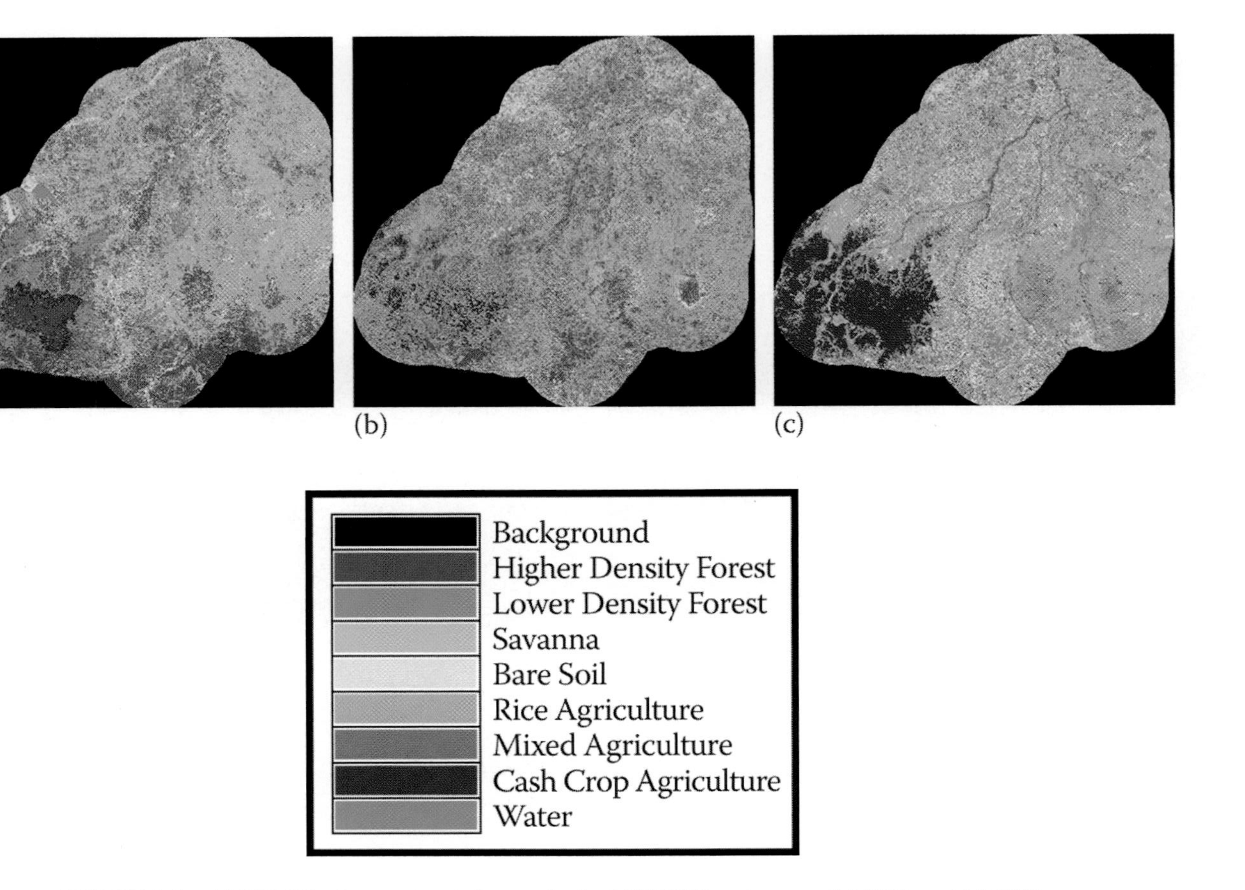

FIGURE 6.3 (a) LULC in the greater study area in the 1972/1973 water year; (b) 1985; and (c) 1997.

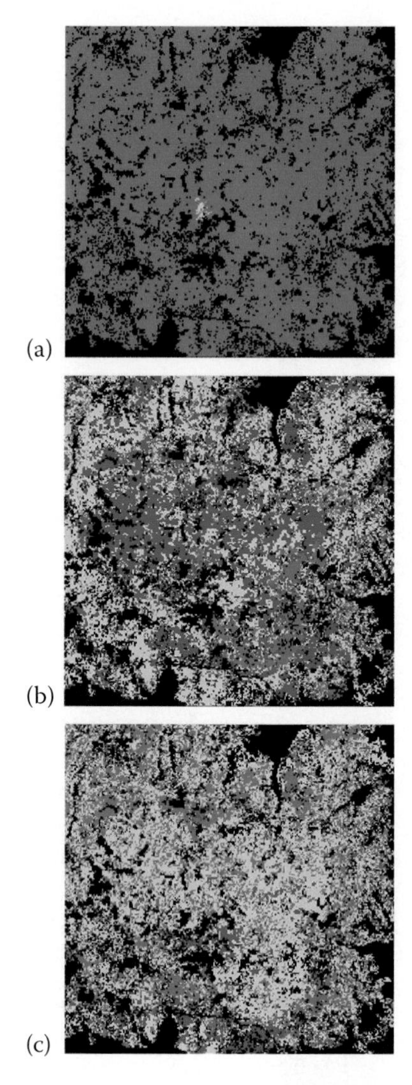

FIGURE 6.6 (a) The change in configuration from 1972/1973 to 1975/1976, revealing that the entire subset area has experienced a greater than 10% increase in interspersion/juxtaposition index (IJI) scores (due to increased fragmentation and concomitant interdigitation). Note that most of the area experienced the same type of change. (b) Illustration of a different trend between 1975/1976 and 1979, whereby increases, decreases, and relative stability in IJI vary spatially. More upland areas (most central in the subset) experienced a consolidation on the landscape, while peripheral areas remain relatively stable in terms of configuration with notable exceptions on the southeastern perimeter. (c) Illustration of the continued spatial heterogeneity in IJI, with lowland/peripheral areas undergoing continued fragmentation, while the less accessible, upland areas appear to have leveled off in terms of larger-scale fragmentation or consolidation but continue to experience small pockets of fragmentation throughout.

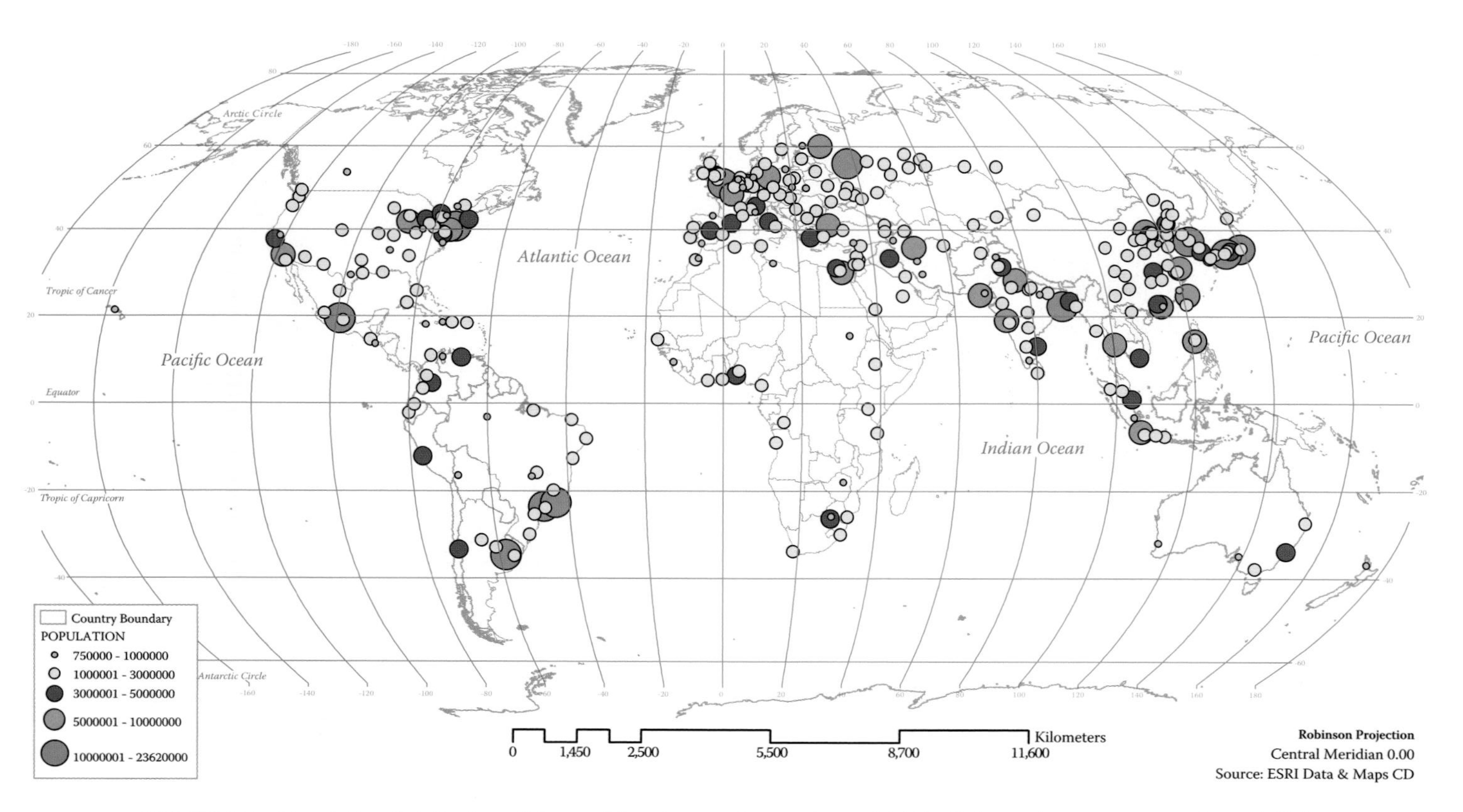

FIGURE 8.1 Population distribution of the world's urban agglomerations with 750,000 people or more around year 2000.

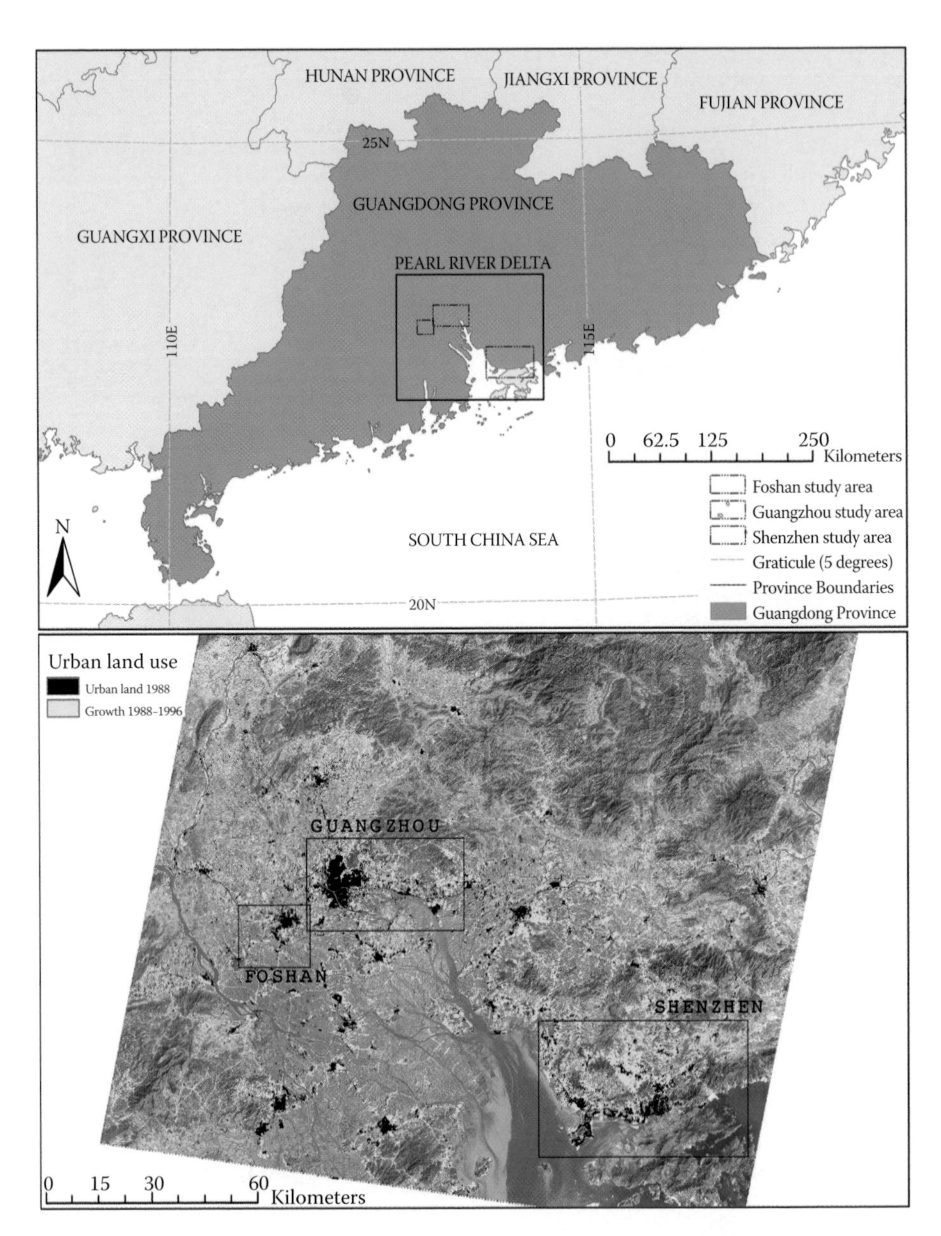

FIGURE 8.2 The Pearl River Delta in southeast China and the Shenzhen, Foshan, and Guangzhou study areas (urban land in 1988, new urban land between 1988 and 1996).

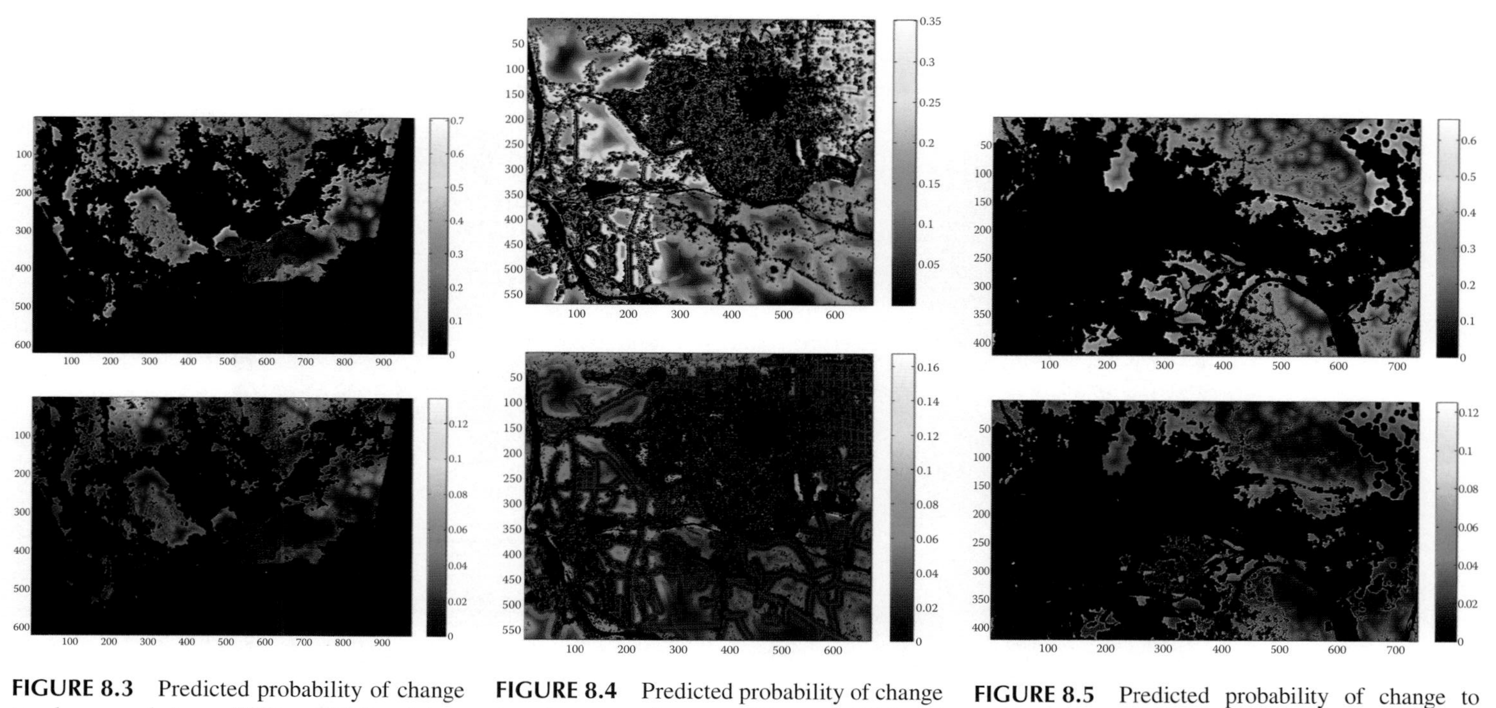

FIGURE 8.3 Predicted probability of change to urban areas between 2004 and 2012 and standard deviation of pixel predicted probabilities for Shenzhen (60 m resolution; values in percentage points).

FIGURE 8.4 Predicted probability of change to urban areas between 2004 and 2012 and standard deviation of pixel predicted probabilities for Foshan (30 m resolution; values in percentage points).

FIGURE 8.5 Predicted probability of change to urban areas between 2004 and 2012 and standard deviation of pixel predicted probabilities for Guangzhou (60 m resolution; values in percentage points).

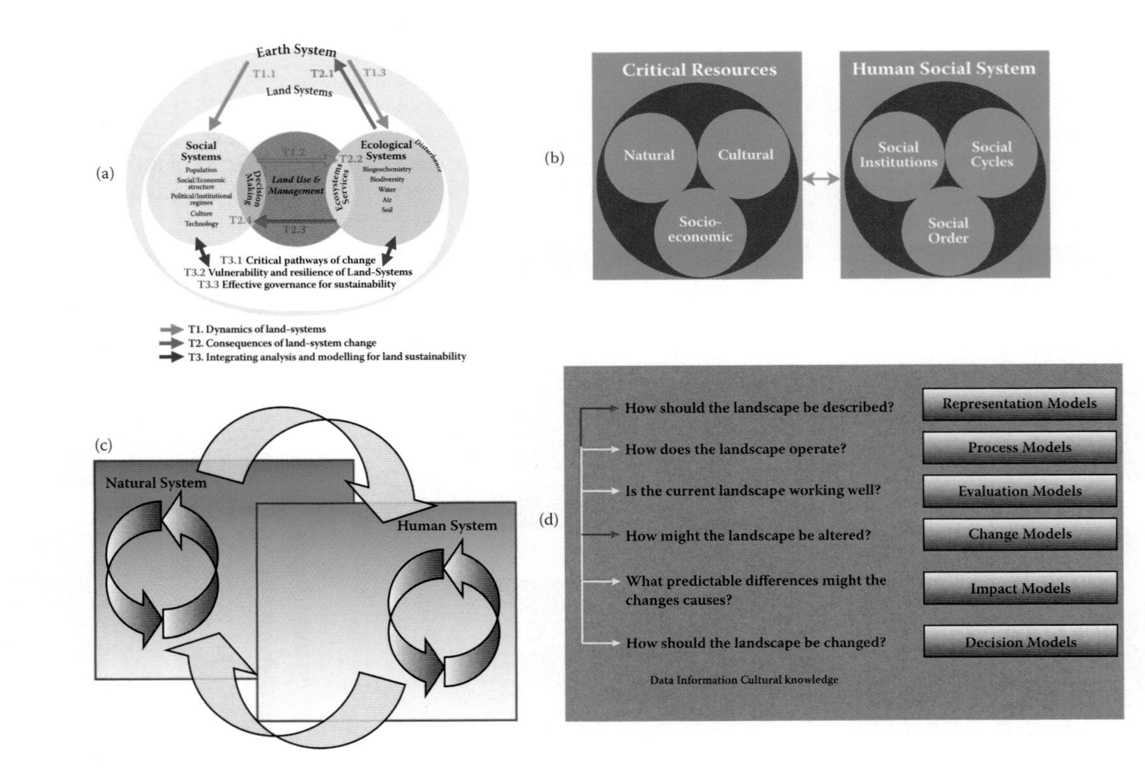

FIGURE 9.1 A variety of integrating frameworks that seek interdisciplinary definition and focus on key questions within the broad scope of management of land use change in coupled human environment systems. (a) Analytical framework for the Global Land Project of IGBP and IHDP. (From GLP, 2005, with permission.) (b) The Human Ecosystem Model. (From Machlis et al.,[8] 1997, with permission.) (c) The U.S. National Science Foundation program on Biocomplexity in the Environment (http://www.nsf.gov/geo/ere/ereweb/fund-biocomplexity.cfm); (d) The Landscape Design Research Framework. (From Steinitz et al.,[9] 2003, with permission.)

3. In the analysis of landscape metrics described in this chapter the results from Trani and Giles's study may diverge from our conceptual model at the plot exhaustion stage in ways that we do not yet understand. This may occur because whereas the deforestation simulated by Trani and Giles was for progressive deforestation, what happens in colonization schemes is that there can be significant regrowth in the medium to later stage in the sequence of deforestation.

7.5 DISCUSSION

We see the conceptual model we have developed from Chapare as a preliminary attempt to thicken our understanding of why particular patterns of fragmentation occur in a so-called fishbone colonization scheme. The regularity of the spatial imprints of roads and property grids in such areas made them interesting candidates for this initial investigation (Figure 7.7). This study is, however, limited, partly by the relatively small number of communities we have researched intensively, which might lead to context specific generalizations,[42] and partly because we have used retrospective analyses of decision making by colonists. While taking a different approach to Perz and Walker[25]—by focusing on colonist's responses to changes in economic conditions and anti-coca policies and the loss of primary forest—we join them in their clarion call for researchers to focus on the forest outcomes of small-farm colonists. They state that *"most land use models ... do not take account of land taken out of production and left to fallow"* (p. 1009), and we would argue that most land use models do not take account of land taken out of production, left to fallow, *or how land under primary and secondary forest is spatially configured.* Only if we research along these lines will we be able to apply methods such as those advocated to plan colonization in forest areas[43] or be able to restore fragmented areas.[44]

The field then is open for further research, and we argue the following lines of investigation are needed to deepen our understanding further:

1. More communities in Chapare could be investigated, particularly those with physical and socioeconomic characteristics other than those outlined in Table 7.1. However, a more profitable line of investigation would be to research clusters of adjacent communities. This would allow interactions between communities at their boundaries to be investigated and might also reveal the extent to which cooperation between communities takes place, or gauge the potential of cooperation in the future.
2. Comparative research between Chapare and similar—fishbone—colonization schemes in the Amazon Basin would enable us to further consider the robustness of many aspects of the conceptual model we have developed and enable us to move toward a general tool that could be used for basin-wide integrated planning in colonization schemes.
3. So far, our model relies on a retrospective analysis of data. However, its utility for planning lies in its ability to integrate conservation in the planning of future colonization in the Amazon Basin. Therefore, research into decision making by individual land owners and *sindicatos* under different

FIGURE 7.7 (a) Chapare road building. The extension of the road network in Chapare continues. This photograph was taken in August 2003, and shows a road being pushed deep into relatively intact lowland tropical forest along the line of a former footpath which linked some isolated settlements to the former road head. (b) In the east of Chapare cattle rearing is the main farming activity. Land parcels here are 50 ha in area, compared to 20 ha in settlements dominated by cultivation. (c) Bananas are one of the main alternative crops to coca in Chapare. They often result in large areas being cleared to compensate for loss of coca income which leads to large areas of nonforest monoculture. The road cutting through this banana plantation is an unimproved primary feeder road.

future economic, political, and environmental scenarios is both attractive to researchers and essential to planners.

4. As we noted in the introduction, this study is grounded in one clearance typology—planned settlement leading to fishbone patterns of land clearance. Other typologies require similar research.

Based on our observations a nagging question remains: is it possible to build structures for conservation in plans for colonization schemes? The requirement is to set aside some land within the colonization schemes that will either act as "sinks" or "reservoirs" with the colonization zone, or to plan wildlife corridors—either contiguous forest corridors or stepping-stones of forest patches for migration across and within the colonization landscapes. Theoretically this is possible, but where does the land come from? In Chapare the original cadastre had land that was not assigned to settlers; often strips of forest along rivers or, to the north, forests that are inundated for many months each year. In other words wildlife corridors and forest patches were "planned by accident" but were not afforded protection until forests along watercourses were specifically protected in the new Forestry Law that was passed in 1996.[45] Disappointingly from the viewpoint of conservation, but not surprisingly given the demands on land and the laissez faire attitude to spontaneous colonization, many of these unallocated lands have been occupied, the exception being the inundated forests. For example, adjacent to Communidad Arequipa a strip of forest land along a river has been illegally colonized, and an unallocated forest area adjacent to Communidad Caracas was added to the community as it expanded after they had petitioned INC.

If we fail to thicken our understanding of how land managers make land use decisions in landscapes of colonization, primary forests will continue to disappear in ways we do not comprehend, secondary forests (which have different ecological properties and conservation values to primary forest) will come and go, and the spatial outcomes may continue to surprise us. These landscapes will lose their ability to allow faunal and floral interchange between "intact forest blocks" as the deforestation fronts they represent close down. New deforestation fronts will open up. If we are unable develop a "thick" understanding and inject it into planning processes, the disease of "thin" understanding will continue to prevail and in all likelihood the scenarios outlined for road corridors in Brazil by Fearnside[29] and Laurence et al.[30,31] will become commonplace in Amazonia.

ACKNOWLEDGMENTS

Part of this research was funded by the European Union (Contract ERBIC189CT80299). Andrew Bradley's PhD research was partly funded by this grant, as well as a Slawson Award from the Royal Geographical Society with the Institute of British Geographers, and funding from the University of Leicester. We are grateful to Richard J. Aspinall and Michael Hill for inviting us to present our research, to Christian Brannstrom for critically commenting on an early version of this chapter, and to Felix Huanca Viraca for accompanying us in the field.

REFERENCES

1. Godron, M., and Forman, R. T. T. *Landscape Ecology*. John Wiley & Sons, New York, 1986.
2. Fahrig, L. Effects of habitat fragmentation on biodiversity. *Annual Review of Ecology and Systematics* 34, 487–515, 2003.
3. Ries, L. et al. Ecological responses to habitat edges: mechanisms, models and variability explained. *Annual Review Ecology, Evolution and Systematics* 51, 491–522, 2004.
4. Turner, I. M. Species loss in fragments of tropical rain forest: A review of the evidence. *Journal of Applied Ecology* 33, 200–209, 1996.

5. Debinski, D. M., and Holt, R. D. A survey and overview of habitat fragmentation experiments. *Conservation Biology* 14, 342–355, 2000.
6. Bierregaard, R. O., Jr., and Lovejoy, T. E. The biological dynamics of tropical rainforest fragments. Bioscience 42, 859–866, 1992.
7. Laurence, W. F., and Bierregaard, R. O. Fragmented tropical forests. *Bulletin of the Ecological Society of America* 77, 34–36, 1996.
8. Laurence, W. F., and Bierregaard, R. O., eds. *Tropical Forest Remnants*. University of Chicago Press, Chicago, 1996.
9. Laurance, W. F. et al. Ecosystem decay of Amazonian forest fragments: A 22-year investigation. *Conservation Biology* 16, 605–618, 2002.
10. Husson, A. et al. Study of forest non-forest interface: Typology of fragmentation of tropical forest. TREES Series B, Research Report No. 2, European Commission EUR 16291 EN, Brussels, 1995.
11. Lambin, E. F. Modelling and monitoring land-cover change processes in tropical regions. *Progress in Physical Geography* 21, 375–393, 1997.
12. Mertens, B. and Lambin, E. F. Spatial modelling of deforestation in Southern Cameroon. Spatial desegregation of diverse deforestation processes. *Applied Geography* 17, 143–162, 1995.
13. Imbernon, J., and Branthomme, A. Characterization of landscape patterns of deforestation in tropical rain forests. *International Journal of Remote Sensing* 22, 1753–1765, 2001.
14. Hargis, C. D., Bissonette, J. A., and David, J. L. Understanding measures of landscape pattern. In: Bissonette, J. A., ed., *Wildlife and Landscape Ecology: Effects of Pattern and Scale*, Springer-Verlag, New York, 231–261, 1997.
15. Rudel, T., and Roper, J. The paths to rain forest destruction: Crossnational patterns of tropical deforestation, 1975–90. *World Development* 25, 53–65, 1997.
16. Geist, H. J., and Lambin, E. F. What drives tropical deforestation? LUCC Report Series No. 4, CICAO, Louvain-la-Neuve, 2001.
17. Lambin, E. J., Geist, H. J., and Lepers, E. Dynamics and land-use and land-cover change in tropical regions. *Annual Review of Environment and Resources* 28, 205–241, 2003.
18. Alves, D. Space-time dynamics of deforestation in Brazilian Amazonia. *International Journal of Remote Sensing* 23, 2903–2908, 2002.
19. Arima, E. Y. et al. Loggers and forest fragmentation: Behavioral models of road building in the Amazon Basin. *Annals of the Association of American Geographers* 95, 525–541, 2005.
20. Brondizio, E. S. et al. The colonist footprint: Toward a conceptual framework of deforestation trajectories among small farmers in frontier Amazonia. In: Wood C., and Porro, R., eds., *Deforestation and Land Use in the Amazon*. University Press of Florida, Gainesville, Fla., 133–161, 2002.
21. Walker, R. T. Mapping process and pattern in the landscape change of the Amazonian frontier. *Annals of the Association of American Geographers* 93, 376–398, 2003.
22. Walsh, S. J. et al. Characterizing and modeling patterns of deforestation and agricultural extensification in the Ecuadorian Amazon. In: Walsh, S. J., and Crews-Meyer, K., eds., *Linking People, Place and Policy*. Kluwer, Boston, 187–215, 2002.
23. Moran, E. F. et al. Integrating Amazonian vegetation, land-use and satellite data. *BioScience* 44, 329–338, 1994.
24. Pichón, F. J. The forest conversion process: a discussion of the sustainability of predominant land uses associated with frontier expansion in the Amazon. *Agriculture and Human Values* 13, 32–51, 1996.
25. Perz, S. G., and Walker, R. T. Household life cycles and secondary forest cover among small farm colonists in the Amazon. *World Development* 30, 1009–1027, 2002.
26. Reid, J. W., and Bowles, I. A. Reducing the impacts of roads on tropical forests. *Environment* 39, 10–17, 1997.

27. Holdsworth, A. R., and Uhl, C. Fire in Amazonian selectively logged rain forest and the potential for fire reduction. *Ecological Applications* 7, 713–725, 1997.
28. Nepstad, D. et al. Road paving, fire regime feedbacks, and the future of Amazon forests. *Forest Ecology and Management* 154, 395–407, 2001.
29. Fearnside, P. M. Avança Brasil: Environmental and social consequences of Brazil's planned infrastructure in Amazonia. *Environmental Management* 30, 735–747, 2002.
30. Laurence, W. F. et al. The future of the Brazilian Amazon. *Science* 291, 438–439, 2001.
31. Laurence, W. F. et al. Deforestation in Amazonia. *Science* 304, 1109–1111, 2004.
32. Nepstad, D. et al. Avança Brasil: *Os Custos Ambientais para Amazônica*. Instituto de Pesquisa Ambiental de Amazônica (IPAM), Belém, Brazil, 24 pp, 2000. (Available at: http://www.ipam.org.br/avanca/politicas.htm).
33. Soares-Filho, B. et al. Simulating the response of land-cover changes to road paving and governance along a major Amazon highway: The Santarém-Cuiabá corridor. *Global Change Biology* 10, 745–764, 2004.
34. Bradley, A. V. Land-Use and Land-Cover Change in the Chapare Region of Bolivia. PhD thesis, University of Leicester, UK, 2005.
35. Henkel, R. The Chapare of Bolivia: A Study of Tropical Agriculture. PhD thesis, University of Wisconsin, Madison, Wisc., 1971.
36. Henkel, R. Coca (*Erythroxylum coca*) cultivation, cocaine production, and biodiversity loss in the Chapare region of Bolivia. In: Churchill, S. P., Balslev, H., Forero, E., et al., eds., *Biodiversity and Conservation of Neotropical Montane Forests*. The New York Botanical Garden, Bronx, N.Y., 551–560, 1995.
37. Millington, A. C., Velez-Liendo, X., and Bradley A. V. Scale dependence in multi-temporal mapping of forest fragmentation in Bolivia: Implications for explaining temporal trends in landscape ecology and applications to biodiversity conservation. *Photogrammetry and Remote Sensing* 57, 289–299, 2003.
38. Steininger, M. C. et al. Tropical deforestation in the Bolivian Amazon. *Environmental Conservation* 28, 127–134, 2002.
39. MACA: Ministerio de Asuntos Campesinos y Agropecuarios. Models of Exploración–Tipo, para siete Subregiones del subtropico humido de Cochabamba. IBTA, Chapare, Bolivia, 1991.
40. Trani, M. K., and Giles R. H., Jr. An analysis of deforestation: Metrics used to describe pattern change. *Forest Ecology and Management* 114, 459–470, 1999.
41. McGarigal, K., and Marks, B. Fragstats. Spatial pattern analysis program for quantifying landscape structure. Forest Science Department, Oregon State University, Corvallis, 1994.
42. Fujisaka, S., and White, D. Pasture or permanent crops after slash-and-burn cultivation? Land-use choice in three Amazon colonies. *Agroforestry Systems* 42, 45–59, 1998.
43. Venema, H. D., Calamai, P. H., and Fieguth, P. Forest structure optimization using evolutionary programming and landscape ecology metrics. *European Journal of Operational Research* 16, 423–439, 2005.
44. Lamb, D. J. et al. Rejoining habitat remnants: Restoring degraded forest lands. In: Laurence, W. F., and Bierregaard, R. O., eds., *Tropical Forest Remnants*. University of Chicago Press, Chicago, 1997.
45. Government of Bolivia. Ley Forestal 1700. La Paz. Bolivia, 1997.

8 Urban Land-Use Change, Models, Uncertainty, and Policymaking in Rapidly Growing Developing World Cities: Evidence from China

Michail Fragkias and Karen C. Seto

CONTENTS

8.1 INTRODUCTION

Projections suggest that as the world's urban population will jump to 61% by 2030 (from today's 50% mark), most of this urban growth will occur primarily in less developed countries, and in Asia in particular.[1] Much interest already exists in megacities—cities with populations of 10 million or more—on which a significant amount of information is being collected. It has been noted though that the majority of urban growth will occur in medium-sized cities.[2] Given that urban growth is a major component of global environmental change[3,4] and the danger of potential undesirable environmental and social effects caused by high rates of growth is ever-present, the relative importance of studying medium-sized cities versus megacities cities in the next century is high. Furthermore, developing world cities have limited human and financial resources employed in various aspects of policy making. Consequently, the collection of reliable data and the use of more advanced methods in planning practice and policymaking becomes extremely difficult (Figure 8.1).

Policymakers in developing world cities are increasingly faced with pressure to assess the impact of their land use strategies and policies[5] as high population growth trends are predicted for at least the next 25 years. Potential socioeconomic and environmental impacts of policies can be assessed with quantitative models. Given the number and underlying motives of different approaches to modeling, policymakers, especially those in developing world cities, could benefit from assistance in choosing the most appropriate model. Is current technology or methodology advancement based on current and recent future realities of medium-sized developing world cities? Pros and cons of different modeling approaches for land use policy making need to be evaluated given the particularities of such cities (e.g., the problem of incomplete and scarce information). The success of sustainable development efforts relies significantly on the identification of better (as accurate as possible) forecasting schemes regarding rates and patterns of future urban development that also connect better with the process of policy making. Thus, this chapter provides an inquiry into questions and tradeoffs a policymaker faces when it comes to the choice of context-specific suitable modeling tools and the establishment of guidelines assisting the decision-making process.

The purpose of this chapter is threefold. First, it discusses issues of urban land use change modeling and explores the intersection of land use modeling with urban policy making at different scales in the context of developing world cities. Second, it discusses the effects of uncertainties in the data sources, theories, and models methods of addressing these issues. Third, it reviews a predictive model of rapid urban transformation that relates a standard modeling tradition to an explicit uncertainty-reducing policy-making framework using Chinese cities as a case study.

8.2 MODELING URBAN LAND USE CHANGE, POLICYMAKING, AND UNCERTAINTY

8.2.1 Modeling Urban Land Use Change

Urban areas, and their form and function, have been studied in the contexts of urban planning, urban economics, urban geography, and urban sociology, much of which are grounded in the spatial land use models of von Thünen.[6] A need for quantitative answers regarding the effects of extent, rate of change, and patterns of global urban land use change has led to the development of urban land use change models (ULCM). A ULCM is a simplification of reality, and its success lies in retaining the fundamental characteristics of the system by simplifying reality as much as necessary (but not beyond that). Thought of as a tool, a ULCM targets "usefulness"; unfortunately, this capacity is not always succinctly stated or demonstrated.

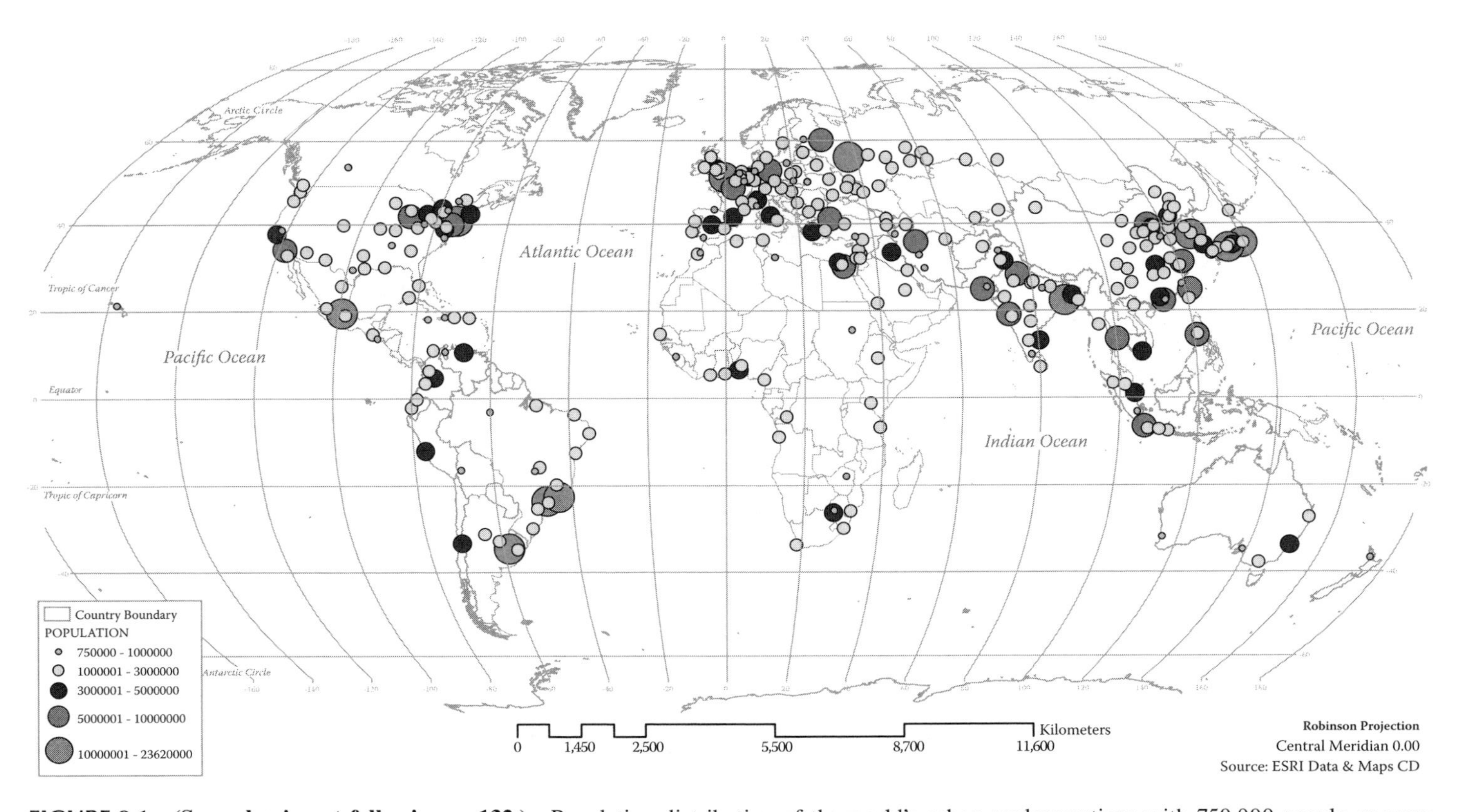

FIGURE 8.1 **(See color insert following p. 132.)** Population distribution of the world's urban agglomerations with 750,000 people or more around year 2000.

The usefulness of an urban land use change model is judged in connection to the goal of the modeling exercise; these goals can be tightly or loosely connected with goals of policymakers. In this chapter we discuss two distinct functions for a ULCM: explanation and prediction/forecasting (that leads to prescription). In its first function, it helps the researcher or the policymaker improve his or her understanding of processes that lead to change and shed light on elements of causality guided by and testing alternative theories of urban growth. In its second function, it can describe and predict the types of land use change that occur (type, amount, rate, pattern, and timing of changes) and more promptly lead to prescription.

There are now dozens of land use models available; a review and typology of (urban) land use change models has been presented in detail elsewhere.[7,8,9,10] Many new "flavors" of modeling are being developed.[11,12] This proliferation reflects the methodological progress in the attempt to understand or predict the nature of the landscape, the types of changes occurring, the causal structure connecting the underlying factors of change, and the hypotheses to be tested. Alternative classifications of urban land use change models include a three-dimensional continuum of spatial scale, time scale, and human decision-making,[11] overlapping categories of equation-based system, statistical technique, expert, evolutionary, cellular, hybrid, and agent-based,[13] and distinct categories of large-scale, rule-based, state-change, and cellular automata.[14]

ULCMs often claim policy relevance but lack a clear definition of the degree of this relevance. Land use change modeling is currently weakly coupled with land use policy making. Although we do not claim a need for a very strong coupling (due to the adverse resource and political reality for such a task in developing countries), we suggest that it needs to be strengthened for optimal knowledge utilization in the policy-making process. This can be achieved by explicitly introducing mechanisms for model uncertainty reduction and a policy-making module in land use change models. It is very important that the relevance of models is more clearly understood and future directions reevaluated. In what follows, we address issues existing at the modeling–policy-making interface.

8.2.2 Policy Making

Policy-relevant land use change models may target a variety of types, levels, and stages of policy-making activity that heavily influences observed land use patterns. Some facets of urban land use change derive in part from policies implemented (synchronously or asynchronously) at different administrative unit levels: at local municipal, county, state, prefecture, and regional levels. National macroeconomic, regional, and local policies have dramatic direct and indirect effects on agents' choices of current and future land use. Policymakers at these levels include a range of public officials, such as urban and environmental planners, and various administrators at local government agencies.

At the local and national levels, concerns regarding social welfare measured in levels of consumption, productive activity, city amenities and disamenities, externalities and ideas of sustainability guide policy-making efforts in targeting—among other goals—an "optimal" urban area size, shape, and population mix. Local urban and exurban governments consistently utilize zoning, growth controls, and taxation/subsidies to drive urban growth and regulate, distribute, or redistribute gains from urban development (while implicitly targeting that the maximization of property values in urban areas). Increasingly, environmental concerns regarding the impact of urban land use conversions also direct policy making. At the global level, institutionally designed policies influence processes of urban land use change in a multitude of ways through the establishment of incentive schemes and structural adjustment programs. Close monitoring of urban population trends suggests the heightened interest of international financial and other institutions (such as the United Nations and the World Bank) that drive global change.

Many theories of policy formation exist, with different assumptions regarding knowledge utilization within the formation process. Most urban growth models are not usually explicit on their assumptions regarding the policy-making process; the most widely adopted view of the policy-making process is that of the rational linear process or agenda setting theory. As with most technical analysis entering the policy realm, the policy relevance of a land use model is of a more informational nature rather than a concrete policy driver nature.

From the rational policymaker's standpoint, the use of a land use change model involves a sequence of decision-making steps and actions that requires (i) the examination of available modeling options, (ii) the choice of model evaluation criteria and their weights (depending on the preferences of the policymaker and the realities of the policy-making setting), (iii) evaluating the model by the selected criteria, and (iv) deriving the overall evaluation through the collection of individual weighed criterion evaluations.[5] Criteria for the selection of a modeling process may include the emphasis on prediction versus explanation, the data sparseness or richness of the policy-making environment, levels of uncertainty in the quality of the data, the emphasis on probabilistic versus heuristic/mechanical approaches, the flexibility of the model to alternative variable specifications, the sophistication in accuracy assessment (validation) of predictions, the need of deep versus basic understanding of the modeling approach, the need for weak versus strong coupling of modeling with the process of policy making, the model's capacity to inform about a variety of policy-making goals and at different levels of policy making, and the emphasis on the theoretical foundation of the modeling approach. Several of these criteria are discussed in more detail below.

8.2.3 The Intersection of Modeling and Policy Making

Recently we have witnessed a scarcity of application of ULCMs for developing world cities. This reflects an underestimation of the potential effects of urban growth in LDCs, the problematic nature of empirical work in LDCs, a lack of understanding of what could be the best modeling option available to a decision maker in a developing world city, and a dearth of applicable ULCMs. Through our work we argue that present pressing predictive needs elevate the importance of statistical models that utilize a minimal input scheme. Models with simple input requirements can find wider application in addressing current and future needs in these cities. Datasets in LDCs are scarce and in many cases inexact due to institutional factors and limited resources. A future increased allocation of resources toward the collection of detailed georeferenced socioeconomic data by the governments of these countries is not certain, and although data are being collected at an international level, this occurs at a very slow pace and at a quite aggregated level. Furthermore, problematic measurement can be catastrophic for the predictive power of models that are successful in capturing the true data-generating process (DGP).

Understanding the importance of knowledge utilization in decision making, the question of the relative importance of explanation versus prediction for policymakers arises. When does a policymaker need (i) predictions regarding the location and timing of land use change under alternative scenarios and/or (ii) the knowledge of whether theoretical hypotheses stand up to statistical tests and of magnitudes of the expected changes associated with shifts in a variety of policy leverages and vice versa? It is not clear if the policymaker always needs a deeper understanding of processes and knowledge of the causation chain. Possibly, the answer to such a question depends on the actual policy-making formation process and the type/level of government or institution responsible for the decision. Various authors suggest that, at a minimum, policymakers should be able to understand the foundations of a modeling approach or at least be able to identify how the results are generated.[10,15] Given the number and level of complexities of alternative modeling approaches, this may be an unrealistic target. Our experience with developing world cities shows that policymakers are definitely more interested in knowing how shifts in policies affect outcomes; they may not want to know the inner workings of the model.

Policymaker preferences over output defines if and when the policymaker has a stake in the choice of methodology (e.g., process-based or mechanistic models). Although socioeconomic processes generate the observed landscape outcomes, models that belong to a rule-based approach may in fact result in better predictions than process-based models utilizing socioeconomic data. This can be partly attributed to data imperfection: variables capturing the socioeconomic processes can be inaccurate or simply these processes might be hard or impossible to quantify. Mechanistic models use data constructs that are based on proximate (rather than underlying) causes of land use change. Unfortunately, these models are also more sensitive to omitted or inexistent information, a fact that can potentially misguide policy making. Process-based models can still be successful to various degrees for forecasting, depending on the geographical location, methods, and aerial unit level of analysis employed for prediction. Such models with proven high predictive power are also usually based on proximate rather than underlying causes. Naturally, successes in predictive capability of rule-based models do not void the search for a DGP.

Models—as opposed to theories—of land use change theory have been more popular tools for policy making and have been developed more for both substantive and practical reasons.[10] Substantive reasons include the complexity of the land use change phenomena and the complex interrelations of various institutional, cultural, political, economic, and social change determinants in theoretical work. Practical reasons include the availability of resources and the "demands of the of the decision making 'clientele'." In short, solid quantitative results that are marketable, visually powerful and ready for use as tools for a wide range of decision makers are valued more highly—and models produce results much better than theory.*

Established approaches in different scientific disciplines and pressures regarding peer acceptance and career advancement also drive methodology-related choices for urban land use change models and are partly responsible for loose connections of models with policy making. The evidence for this is anecdotal, as the authors have been exposed to such complaints in personal discussions with other researchers. In short, the producer's (an academic researcher) incentives for considerable output in the form of journal publications may lead to models loosely or vaguely connected to the practice of policy making. This issue is admittedly difficult to resolve under the current practices.

Awareness of the theoretical foundation of an approach may also be important for the policymaker's choice. Urban cellular automata (CA) models, for example, have a theoretical grounding on ideas of cities as self-organized and emergent phenomena in bottom-up complex systems and fail to capture urban growth in the top-down political dimension. Unfortunately, these are still "largely abstract arguments."[16] Even cutting-edge advanced multiagent system CA (MAS/CA) simulating cities "at the fine scale using cells, agents, and networks" are for now far from being ready for any practical use or "largely ... pedagogic" value.[16]

As the decision-making clientele of urban land use change models targets a variety of goals, a good model should accommodate such a variety. Policymakers care for different size administrative areas depending on whether they are employed at a local, provincial, or national level. Models should be able to address needs of each level of decision making and distill results derived at the highest level to the lowest level and vice versa. Connected to this issue is the capacity to address single or multiple neighboring urban areas in the same model. Single urban metropolitan area analysis is not inclusive of surrounding regional spatial dynamics (immigration and out-migration flows, trade flows, and so forth), an important limitation since cities are interconnected nodes within a network of flows, as well as components of a system of central (urban) places.

When models are weakly informed by theory, an advantage of a ULUC model is its flexibility in allowing the user to make decisions on model specification (although the danger of model selection is ever present—this is addressed in the next section). Current design of mechanical rule-based models shows some inflexibility to alternative specifications, with a resulting "one size fits all"/"cookie-cutter" feel. Finally, an important consideration is the limitations in spatial representations of alternative scenarios imposed by the ULCM, assuming that quantifiable information on potential alternative directions in local, regional, or national policies can be provided. Policies such as zoning or growth controls are the easiest to represent, while openness to in-migration or other economic information such as market conditions may not be easily incorporated into models. Highly stylized (input-restricted) models are only able to incorporate policies reflecting road development and "off-limits to development" zoning; this is a limiting factor in the capacity of the models to include other forms of policy making.

* Pre-1981 literature on the politics of model use in decision and policy making is reviewed in Briassoulis[10] (1999, chap. 5).

8.2.4 Policy Evaluation and Uncertainties

Model and expert knowledge utilization targets the reduction in the uncertainty of outcome predictions and the consequent effects of these outcomes. Statistical decision theory provides quantitative tools for the reduction of uncertainty in optimal policymaking.[17] Given that any urban policy simulation results are dependent on models, how is the "best" model defined and how is it chosen among all possible models? A model is a single representation of reality, and, although statistical criteria can be used to identify the "best" one, it represents just one of many possible data generating processes; thus, model selection should be avoided. Model selection has been criticized as being a weak basis for policy evaluation and derivation of future prescriptions; the search for a single best predictive model is misguided when it comes to policy-relevant models. Robustness across models, on the other hand, is being advanced for policy-relevant work. Unfortunately, model selection ignores the fundamental dimension of "model uncertainty," but methodologies for robustness of the policy prescription across alternative model specifications can be alternatively utilized.

Methodologies addressing the issues of theory and model uncertainty are now available for incorporation in policy-relevant research.[17,18] Uncertainty over competing theories results from uncertainty about which theory of urban growth should be utilized due to institutional and cultural factors affecting land markets in developing countries or differing assumptions regarding agent decision making; it can lead to models that are not well informed by theory. Model uncertainty results from uncertainty over functional form specification for statistical models and is due to subjective perceived relevance and endogeneity issues as well as the question of appropriate spatial and time lags, proxy variables, and so forth. Incomplete knowledge regarding the best model of a system should force the researcher to explore the sensitivity of modeling approaches to alternative specifications.[18] This framework partially solves the problems of subjectivity and ad hoc specifications in uncertain environments regarding the capacity of a variety of proxies to capture the effect of variables entering the data generating process.

An applied statistical framework of policy-relevant urban growth modeling that accounts for model uncertainty makes an explicit reference to a policymaker (PM hereafter) who examines a set of urban–growth-related policies P for administrative units (e.g., sub-city districts, cities, counties, provinces) through the selection of a single policy p.[17,18] The PM utilizes data d about a metropolitan area's land-use and transportation systems (realizations of a process), and the choice is conditional on a model m of the urban economy (m can constitute alternative theories and statistical specifications). The PM minimizes the expected value of an objective (loss) function $l(p,\theta)$ where θ is a the exogenous state of nature (not controlled by the PM but affecting the influence of p on the loss, l)*.

* Consider X, a vector of targeted urban growth rates per administrative unit per time period defined by the PM. Different policies will drive different rates and patterns of urban growth. The deviation of these growth rates from the targeted growth rates can be expressed as a loss function. Each administrative unit is weighted according to the preferences of the PM. The weights are assigned according to the importance of convergence to the target for each administrative unit: the greater the importance of meeting the target, the higher the weight[18].

Unknown exogenous factors that influence land use decisions (the state of nature in a decision-theoretic framework) such as monetary or fiscal macroeconomic policies lead to the minimization of the expected value of the loss function by the PM. Usually probabilities of the states of nature are conditioned on existing data and selected models: $\mu(\theta|d,m)$. Accounting for model uncertainty, they are not conditioned on m since the PM understands that there is probably no "best" model for the urban land use system. The probability density function (pdf) for θ, $\mu(\theta|d)$, is assumed conditional on the existing data d only (and not on m). With a well-defined loss function and *pdf*, the optimal policy is the one that minimizes the expected loss—the solution to the optimization problem. We thus argue that a ULCM should generate a probability distribution of estimates and predictions as well as their distribution characteristics/properties for each pixel for each time of change. A mathematical formulation of the model can be found in Fragkias and Seto[18] and Brock et al.[17] The following section describes the workings of an urban growth model that takes into account the above considerations.

8.3 A New Approach in Modeling Urban Growth in Data Sparse Environments

8.3.1 Mechanics of the Model

We present here an approach in land use change modeling that can explicitly evaluate policies under uncertainties within a spatial socioeconomic environment by incorporating a methodology that addresses issues of theory and specification uncertainty. The model proposed by Fragkias and Seto[18] is a hybrid spatially explicit model of urban land use change with a foundation on economic and statistical discrete choice models of land use change,[8] adjusted for use in data-sparse environments.

The model first reads the input data available provided through remote sensing analysis and other sources at a defined spatial resolution. In its current implementation these are: urban/non-urban land-use maps, a transportation network, areas excluded from development, and a central business district location. It processes this data producing new images/matrices such as new urban growth between examined years (a binary image from which we extract information on a dependent variable) and a collection of new images from which the model extracts independent variables.*

Next, the model employs statistical analysis for the calibration stage. Separating the study area into two parts (its East and West half), it performs a random sampling of developable pixels of the initial East half of the urban/non-urban image (at time $t0$). It creates a calibration dataset with a single dependent variable and multiple sets of independent variables ready for regression analysis. Using two ($t0$, $t1$) calibration images, the model runs multiple logistic regressions with binary dependent variable y as 'change to urban or no change' between time periods $t0$ and $t1$. The

* In this model we focus on prediction of changes and do not utilize data on socioeconomic processes that result in land use patterns. In an effort to develop a model with minimal data requirements we use spatial density and distance variables to check the predictive accuracy of a model. The model also utilizes district dummy variables, each representing one (or a collection) of the districts of each urban area in the study. Fragkias and Seto[18] presents a more detailed description of the model.

modeling approach presented in this case study systematically incorporates a variety of models (or specifications) and accounts for model uncertainty in land use change related policymaking*. The multiple specification model runs reflect the needs of the employed model averaging technique. For n explanatory variables that can be selected for the models, a total of 2^n sets of alternative specification exist and are utilized in the analysis as alternative regressor sets.

Pseudo-Bayesian model averaging is then performed using the calibration sample. Each 2^n logit model run generates predicted probabilities of change (fitted values of the dependent variable) for each sample point, and a weighted average of the predicted probabilities is calculated; the 2^n sets of fitted values are weighted by their respective (normalized) pseudo-R^2 statistic†. A series of binary sample sets of predicted urban/non-urban land are then created utilizing an array of probability cut-off points (threshold values) that range from 0 to 1. The model compares the series of predicted urban/non-urban values with the actual realization of land use during the time period under study and selects the "optimal" threshold level for the calibration period (the cut-off point that generates the minimum difference between predicted urban land and actual urban land).‡

Model validation occurs at two spatial scales: the individual pixel (through PCP validation) and a chosen administrative unit level (by aggregating pixel level information and validation through sample enumeration). The validation sample is derived through a second random spatial sample within the second (West) half of the study area. All 2^n sets of independent variable values are extracted for the new validation sample. Together with the estimated sets of variable coefficients from the calibration stage they are applied to the fitted probability logit formula for each model. This generates predicted probabilities of change for the sampled developable pixels for time period $t1$ utilizing the average of the fitted/predicted probabilities of all the models (weighted by the normalized pseudo-R^2 score that each model achieves).

The first type of validation occurs through the goodness-of-fit measure of "percent correctly predicted" (PCP)§. Typically, in PCP validation, choice (or prediction) is defined as the alternative with the highest predicted probability. These classifications are then compared with actual changes and the PCP measure is calculated. Apart from the intuitive binary cut-off value of 0.5, any probability threshold value can be set for the generation of binary predicted change values. We automate the selection of this threshold in the calibration stage utilizing the criterion of "best growth rate

* Probabilistic models are sensitive to the problems of predictive bias and lack of calibration: predictive bias is a problem of balance or "*the systematic tendency to predict on the low side or the high side*" (p. 391)[19], and averaging models with alternative specifications increases the chances of averaging out the problem; lack of calibration, is "*a systematic tendency to over- or understate predictive accuracy*" (p.391)[19] and is also a negative factor for validation through thresholding due to an increased sensitivity to it.

† The standard deviations of the predicted probability of change estimates are also calculated (but only at the stage of the full image application).

‡ This is a form of an external imposition of an urban growth rate scenario on the model. A threshold can also be selected in such a way that an alternative urban growth scenario is portrayed.

§ PCP validation occurs in the form of separation by space within an out-of-sample modeling framework.[20]

matching."* This probability threshold value is applied in the new validation sample, the classification is performed, and the PCP measure is calculated.

The second type of validation occurs at a larger scale than the individual pixel through sample enumeration. The technique of sample enumeration sums predicted probabilities over a set of agents or observations with the goal of generatingconsistent estimations of aggregate outcomes.[21] The model utilizes a spatially explicit version of this aggregation method for the purposes of validation and forecasting, summing up probabilities to the district level. It employs sample enumeration for the validation dataset and compares the aggregate estimate of urban change to the actual aggregate change in the sample. Predictive accuracy is defined by the success of the summed averaged predicted probabilities in accurately capturing aggregate change at the selected administrative unit†.

The model predicts land-use in two ways. First, we threshold the predicted probabilities. Prediction results are presented initially for the population of developable pixels in *t*2 and potentially for any discrete number of future iterations (*t*3, *t*4, and so forth). Each iteration accounts for the passing of a single—equivalent to the logit models—time period. The predictions utilize the estimated sets of regression coefficients and (potentially for each iteration) a partially new set of independent variables (reflecting landscape evolution). The optimal calibration threshold is applied again for the "translation" from probability to predicted change. Second, prediction occurs through the sample enumeration technique for forecasting at aggregate administrative unit levels utilizing a hypothetical scenario dataset.

A multiple scenario examination is usually an integral part of a policy decision-making process. Given the nature of the currently incorporated variables, a simulation module can be used for the examination of policy-relevant alternative scenarios regarding simple policy leverages: first, the researcher can define new areas of undeveloped land that are excluded from development; second, altered transportation routes can be designed according to existing plans of road or railway expansion; third, the user can create patches of new developed land of high intentionality (e.g., a new airport) capturing spill-over effects of such developments that would otherwise be difficult to predict. The user/decision maker can feed the model a collection of scenario images and define a loss function that connects predictions of urban growth with the objective the decision maker is trying to achieve (e.g., minimization of agricultural land loss). The model can provide predictions of change and associated

* This automated calibration process though has a disadvantage. Since it is based on an ad hoc criterion selected by the researcher, the selection of a probability threshold over 0.5 is subjective.

† The obvious value of PCP validation is the ease of validation at the pixel level and any aggregated level of groups of pixels (for example, within administrative boundaries). Unfortunately, the application of probability thresholds in statistical models is counter to the notion of a predicted probability in such models. Limited information due to the unobservable component in the formulation of the choice process forbids the prediction of the choice of an alternative for a single unit of observation. The nature of the predicted probabilities in discrete choice models has a standard statistical "large sample repetition" interpretation: the alternative with the highest probability is not the unit's choice each time. This interpretation makes a model design that attempts a cross from probability space to choice space a more complex and difficult task.[21] Thus, PCP measures may not provide the best way to validate a statistical model.

losses for each scenario, thus giving the decision maker the choice of the policy that minimizes the loss accounting for model uncertainty.

8.3.2 Application to Three Cities of the Pearl River Delta, China

While urban land use change is a worldwide phenomenon, it is most dynamic in Asia—and particularly in China—where unprecedented rates of urban growth have occurred over the last two and a half decades.[1] The interaction of compelling local, regional, and global factors has resulted in remarkably varied urban configurations.[22] Our study area is comprised of three of the most developed cities in the Pearl River delta (PRD) region in the coastal southeast China, Shenzhen, Guangzhou, and Foshan (Figures 8.2 and 8.3); these and other cities in the PRD have experienced dramatic urban land growth rates in the past two decades. During an 11-year period—from 1988 to 1999—urban land grew* in the PRD by 451.6% or approximately 16.5% a year[23]. The Delta generates more than 70% of the provincial GDP and is home to 21 million people, nearly one-third of the province's official population.[23] Previous research shows that the region is undergoing rapid urban transformation.[24] Although the cities are in close proximity, they differ in history, demographics, and economics.[18,25,26]

The model is run for the three cities at two pixel resolutions, 30 m and 60 m. Using a random calibration sample from the developable pixels of the initial 1988 urban/nonurban image† the model explores all potential combinations of logit models created by allowing five variables to enter alternative specifications, which results in 32 models.‡ After deriving all the sets of estimated coefficients and calculating the model-averaged predicted probabilities, the calibration stage identifies the thresholds that best fit the observed urban growth rate for the period 1988 to 1996 for all cities.

Validation occurs with the use of the second random sample and the model-averaged predicted probabilities for all cities and specified resolutions. In a case of two discrete states of land use (urban and nonurban), one finds two cases where predictions are accurate (correctly predicted urban and nonurban) and two cases where predictions are wrong (wrongly predicted urban and nonurban). In the case of PCP validation the total percentage of wrongly predicted pixels ranges between 23% 27%. The model consistently generates a disproportionately larger percentage of wrongly predicted nonurban pixels relative to the percentage of wrongly predicted urban pixels, a manifestation of the predictive bias problem: we observe a systematic tendency of smaller probability values. The model also reports comparisons between the aggregate—at the district level—actual change, the aggregate predicted change through sample enumeration, and the aggregate predicted change through thresholding for all cities and resolutions. The sample enumeration technique performs consistently better in the validation of the aggregate counts of change. Across all cities, we observe that the vector distances between the aggregate actual change with aggregate sample enumeration predicted change and the aggregate thresholding

* In particular, the metropolitan areas of cities such as Shenzhen, Guangzhou, and Foshan grew by 132.3% (8% annually), 247.6% (12% annually), and 140% (8.3% annually), respectively

† For the 30 m resolution, this number is close to 14,000 pixels in the Shenzhen study area, 3,300 pixels in the Foshan study area, and 10,200 pixels in the Guangzhou study area (Figure 8.3).

‡ Excluding district dummy variables would provide an additional 32 models.

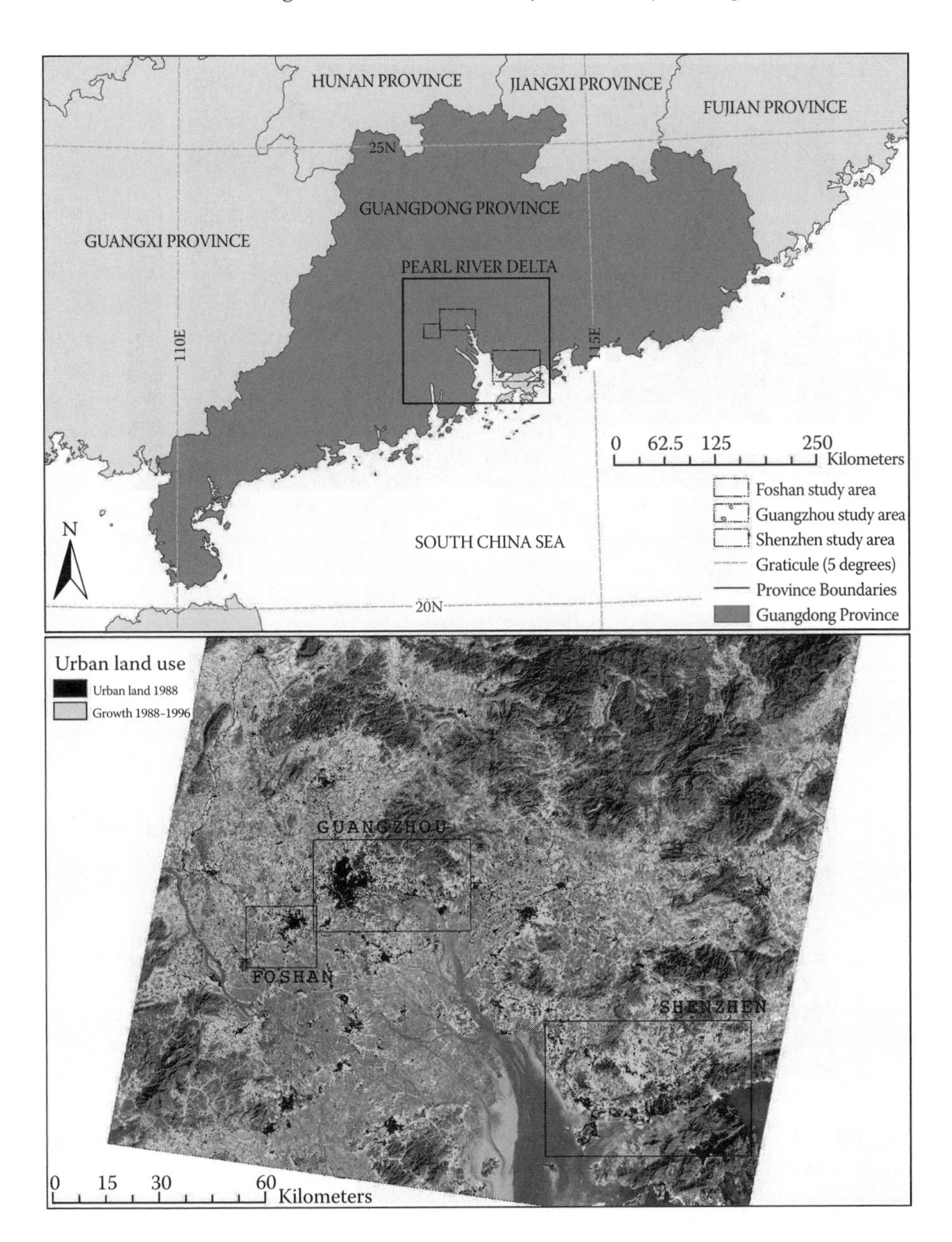

FIGURE 8.2 **(See color insert following p. 132.)** The Pearl River Delta in southeast China and the Shenzhen, Foshan, and Guangzhou study areas (urban land in 1988, new urban land between 1988 and 1996).

predicted change drop significantly as we lowered the resolution from 30 m to 60 m. The performance of aggregate prediction through thresholding improves relative to aggregate prediction through sample enumeration but is still not satisfactory. In short, there exists a clear failure of spatially accurate prediction when using threshold probabilities for transition to choices. The pixel level predictions can be successfully

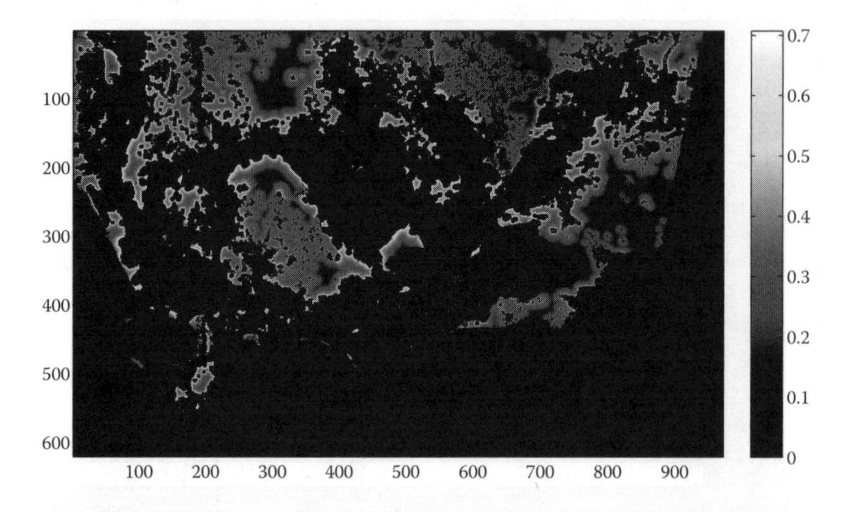

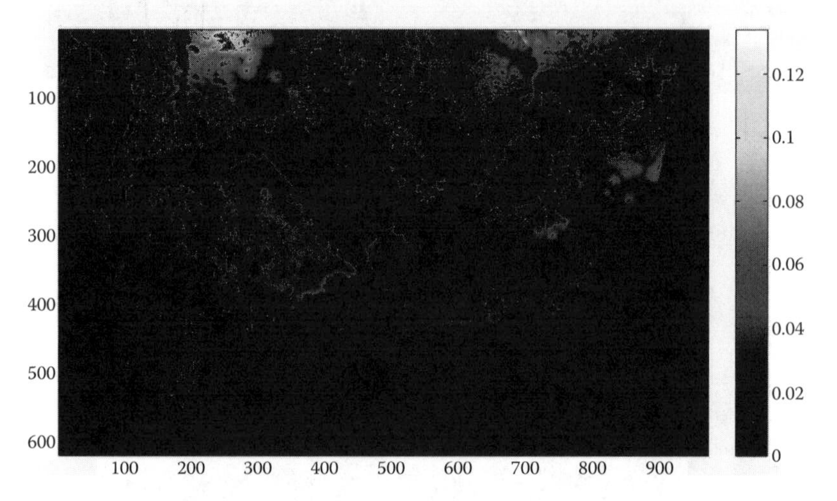

FIGURE 8.3 **(See color insert following p. 132.)** Predicted probability of change to urban areas between 2004 and 2012 and standard deviation of pixel predicted probabilities for Shenzhen (60 m resolution; values in percentage points).

used through aggregation to a larger administrative unit such as the neighborhood or the township through the process of sample enumeration.

At the prediction stage, the model averaging process that generates the predicted probability values is repeated for the full population of observations; the *t*1 period (1996) data provide the values of independent variables, and a full image of predicted probabilities of development is generated. Each pixel is assigned the averaged predicted value from the logit formula using the sets of estimated coefficients. Figures 8.4, 8.5, and 8.6 map these predicted probabilities of development between 2004 and 20012 and the associated standard deviations of the predictions.

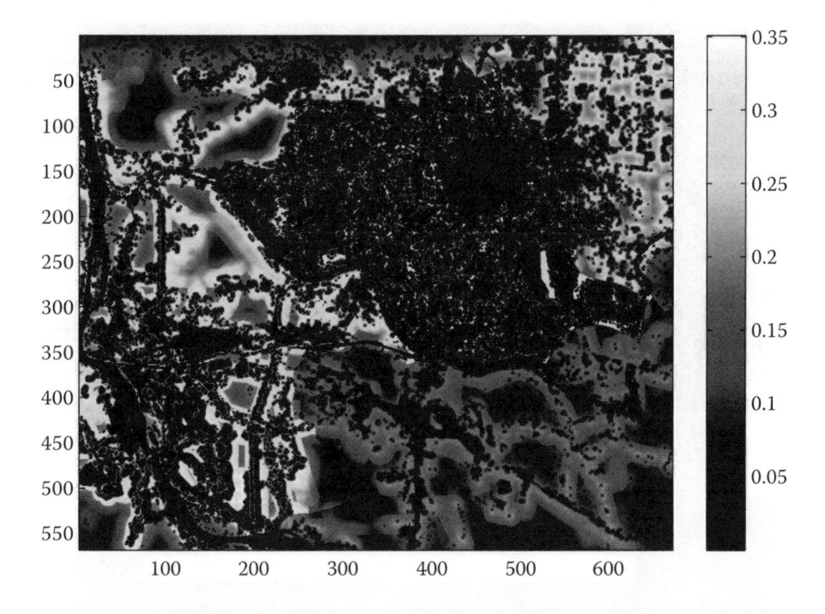

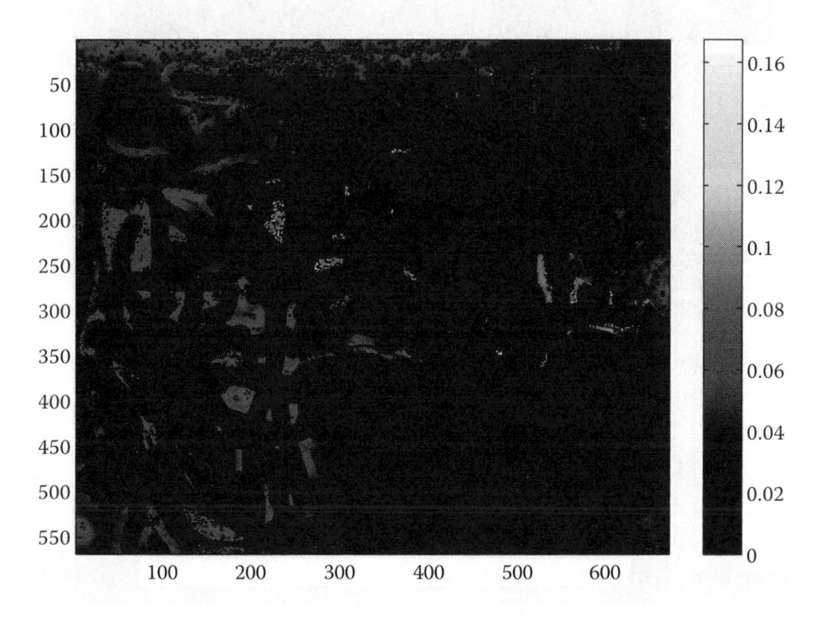

FIGURE 8.4 **(See color insert following p. 132.)** Predicted probability of change to urban areas between 2004 and 2012 and standard deviation of pixel predicted probabilities for Foshan (30 m resolution; values in percentage points).

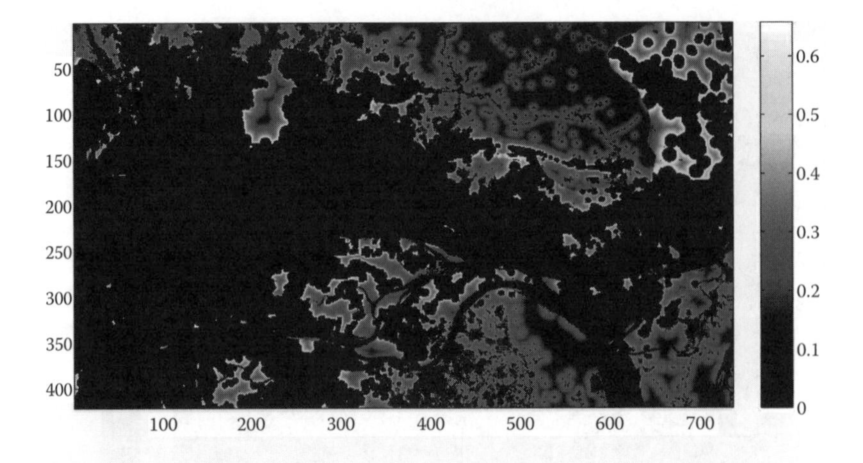

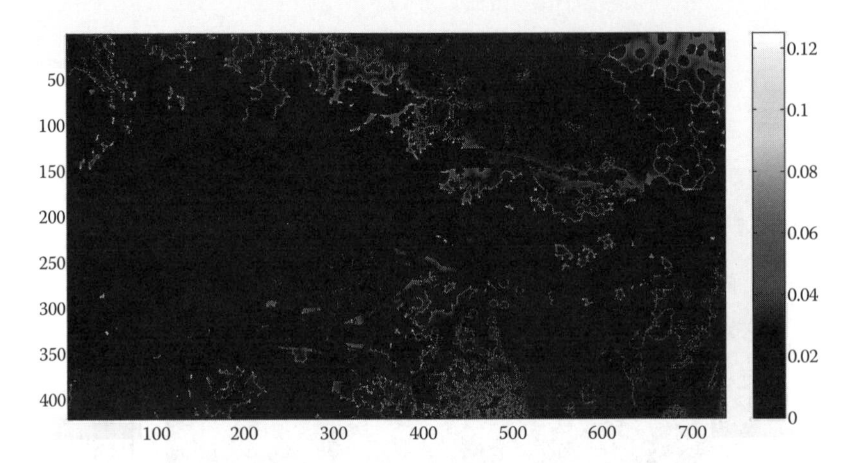

FIGURE 8.5 **(See color insert following p. 132.)** Predicted probability of change to urban areas between 2004 and 2012 and standard deviation of pixel predicted probabilities for Guangzhou (60 m resolution; values in percentage points).

For predictions based on thresholding, the model generates images of urban/nonurban land use for any future time period; Figure 8.7 plots the image for 2012. Amounts of urban land use change for each district between any set of years are predicted through the technique of sample enumeration.

8.4 Discussion and Conclusions

Accounting for model uncertainty in land use change related policy making through the employed methodology reduces uncertainties for a decision maker that are inherent in the use of models in the decision making process. The proposed methodology

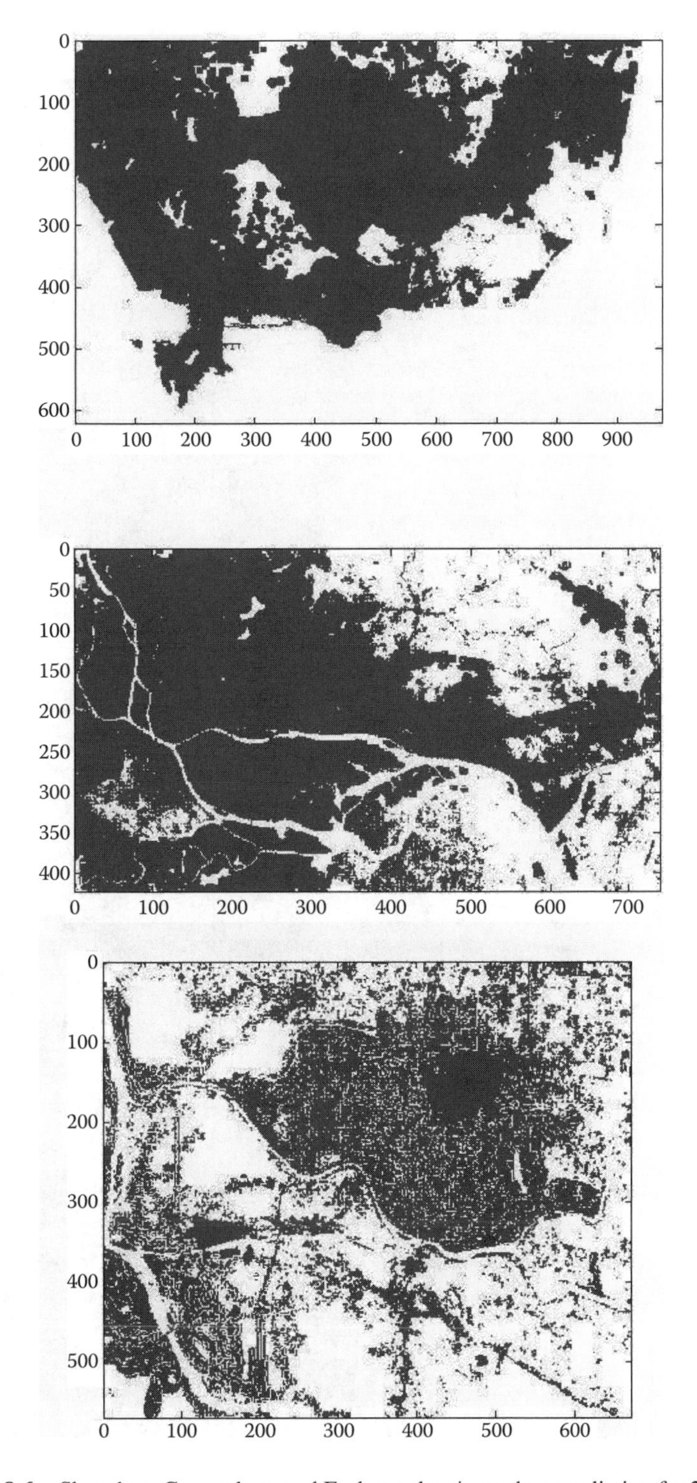

FIGURE 8.6 Shenzhen, Guangzhou, and Foshan urban/nonurban prediction for 2012.

FIGURE 8.7 City of Shenzhen – March 2002.

potentially reduces the problems of predictive bias and lack of calibration. Through the above methodology, a PM is able to calculate a variety of characteristics of probability distributions of outcomes (such as the mean and variance) that may affect policy-making choices. Furthermore, since the methodology enhances the capacity for convergence of the true value of underlying parameters and the estimates of those parameters, a researcher may worry less about possible statistical model biases and can feel more secure in the decreased capacity of manipulation of the model by policy stakeholders toward particular answers. Unfortunately, though, moving from this simple single policy leverage model to multiple simultaneous leverage level choices by PMs increases significantly the complexity of the model. Furthermore, this rational decision-making framework is also far from a perfect depiction of the actual policy-making process; it ignores issues such as the conflicts over ideology, power differentials, problems in communication and collaboration, and the influence of bureaucracies and special interest groups.

The concept of uncertainty reduction through mechanism/model design—and the relevance of this methodology—is supported by modern political economy. Since a realistic view of government supports that politicians, bureaucrats, and policymakers within national and local governments are a mix of actors with public-good and private-gain motivation, policies should be designed with the assumption that all policymakers operate under the homo economicus model (that is, personal gain motivation drives action and is a huge factor in any decision-making process[27]). Given that models and model selection can be employed for the support of private interests, a modeling approach that provides a defense against model selection is desirable.

The model input accuracy issue ("junk in—junk out") naturally affects land-use change models too. Socioeconomic data can be nonexistent, incomplete, inaccurate, unreliable, or all the above in a developing world setting. Even data accuracies of a fundamental variable such as urbanization and its forecasts are severely criticized.[28] Results of modeling approaches that are sensitive to omitted or nonexistent information are potentially misguiding for policy making. We find that enriching datasets with remotely sensed information, in combination with models of a statistical nature, is a positive step in any modeling exercise regarding developing world cities. On the one hand, given data inaccuracies in socioeconomic data, statistical models prove to be less sensitive to input data imperfections, since they assign errors in measurement in the stochastic part of the model and produce unbiased estimates even when errors plague the data (if those errors are not systematic—randomly distributed). On the other hand, the uncertainty over the accuracy of inputs can be controlled/measured more easily for data derived from satellite imagery rather that socioeconomic census sources (assuming the minimization of uncertainty and no bias in satellite imagery classification).

The application of probability thresholds in statistical models is problematic due to the notion of a predicted probability in probabilistic models; pixel-level validation techniques may be misguiding when used in statistical models. The standard statistical "large sample repetition" interpretation of probability (which applies to our statistical model) conflicts with the idea of PCP measures of goodness-of-fit due to this fundamental characteristic of probabilistic discrete choice models. An improvement to standard thresholding techniques (which also automatically discard information provided by the neighboring pixels' predicted probabilities) should be offered by the research community and is a research goal for the authors of this chapter. We suggest that until a more advanced validation mechanism is proposed, statistical models should be effectively limited to mainly the nonspatially explicit sample enumeration validation and prediction. These estimates could potentially be coupled with a more adequate module operating as an pixel allocation mechanism of predicted growth.

The aforementioned problem of predictive bias manifests itself in the model through an imbalance in the percentages of wrongly predicted nonurban/urban pixels. The model presented here partially corrects for this limitation through pseudo-Bayesian model averaging. Still, we find that a significantly limiting characteristic of policy-relevant land use change models is that they do not allow the user to define which type of error is more important.* Future versions of urban growth models should incorporate a mechanism that would allow a PM to control the amount of different types of errors (optimally, by policy region), when the relative costs of wrong predictions by models are significantly different, and thus attain higher policy relevance.

We identify several key points from the above analysis. First, the user of probabilistic choice models should primarily focus his or her attention on the predicted probability maps so that probable hotspots for development are identified rather than generating predicted maps of choices through simple thresholding mechanisms (change or not change in a binary setting). Second, the advantage of utilization of a variety of models is manifested in the fact that maps of standard deviations for the predicted probability maps convey important information regarding the spatially explicit agreement of the various models. Third, in light of scarce information and difficulty of dataset enrichment, predicted probability maps should be coupled with additional information about development trends that may be derived from surveys with important stakeholders in land use change decision making and local knowledge.

Generally, similarly to other models in its class, the success of the model is constrained by the availability of quantifiable proxies. Similar to other models of this type, scenario building requires the use of variables for which future values can be provided. Future implementations of the model will consider (i) validation through separation by time; (ii) the more immediate incorporation of calibration schemes to different localities/districts that calibrate the model and capture in more detail the urban growth rates (the quantity of change) at the district level; (iii) the coupling of this model with other models operating at other aggregation levels, shedding more light into the realistic prospects of development of any model-identified hotspot; and (iv) the collection and analysis of scenarios and policymaker preferences in collaboration with Chinese city planning departments.

* In statistics, for example, when testing a hypothesis about a population, the researcher can control the probability of two distinct types of errors, type I and. type II errors, based on the severity of a possible mistake. The dilemma of whether to adopt a high or low significance level is stronger when the stakes are higher.

REFERENCES

1. United Nations. World Urbanization Prospects: The 2003 Revision. UN Press, New York, 2004.
2. UNCHS. The State of the World's Cities Report 2001. United Nations Centre for Human Settlements (Habitat), New York, 2002.
3. Lambin, E. F. et al. The causes of land-use and land-cover change: Moving beyond the myths. Global Environmental Change 11, 261, 2001.
4. National Research Council. Human Dimensions of Global Environmental Change: Global Environmental Change: Research Pathways for the Next Decade. National Academy Press, Washington, D.C., 293 pp, 1999.
5. Boulanger, P.-M. and Bréchet, T. Models for policy-making in sustainable development: The state of the art and perspectives for research. Ecological Economics. 55, 337, 2005.
6. von Thünen, J. H. Der isolierte staat in beziehung auf landwirtschaft und nationalökonomie. Gustav Fisher, Stuttgart; translation by C. M. Wartenburg (1966), The Isolated State. Oxford University Press, Oxford, U.K., 1826.
7. Brown, D. G., et al. Modeling land use and land cover change. In: Gutman, G. et al., eds., Land Change Science: Observing, Monitoring, and Understanding Trajectories of Change on the Earth's Surface. Kluwer Academic Publishers, Dordrecht, Netherlands, 395, 2004.
8. Irwin, E. G., and Geoghegan, J. Theory, data, methods: Developing spatially explicit economic models of land use change. Agriculture Ecosystems and Environment 85, 7, 2001.
9. EPA. Projecting Land-Use Change: A Summary of Models for Assessing the Effects of Community Growth and Change on Land-Use Patterns. U.S. Environmental Protection Agency, Office of Research and Development, Cincinnati, Ohio, 2000.
10. Briassoulis, H. Analysis of land use change: Theoretical and modeling approaches. In: Loveridge S., ed., *The Web Book of Regional Science*. Regional Research Institute, West Virginia University, Morgantown, W.Va., 2000 (Available at: http://www.rri.wvu.edu/regscweb.htm).
11. Agarwal, C. et al. *A Review and Assessment of Land-Use Change Models: Dynamics of Space, Time, and Human Choice*. U.S. Department of Agriculture Forest Service, Northeastern Forest Research Station, UFS Technical Report NE-297. Burlington, Vt., 2002.
12. Brown, D. G. et al. Path dependence and the validation of agent-based spatial models of land use. *International Journal of Geographic Information Systems* 19, 153, 2005.
13. Parker, D. C. et al. Multi-agent systems for the simulation of land-use and land-cover change: A review. *Annals of the Association of American Geographers* 93, 314, 2003.
14. Klosterman, R. E., and Pettit, C. J. Guest editorial. *Environment and Planning B: Planning and Design* 32, 477, 2005.
15. Szanton, P. *Not Well Advised*. Russell Foundation and the Ford Foundation, New York, 1981.
16. Batty, M. Editorial. *Environment and Planning B: Planning and Design* 31, 326, 2004.
17. Brock, W. A., Durlauf, S. N., and West, K. D.. Policy evaluation in uncertain economic environments. *Brookings Papers on Economic Activity* 1, 235, 2003.
18. Fragkias, M., and Seto, K. C. Modeling urban growth in data sparse environments: A new approach. *Environment and Planning B: Planning and Design* 34, 858, 2007.
19. Hoeting, J. A. et al. Bayesian model averaging: A tutorial. *Statistical Science* 14, 382, 1999.

20. Pontius, R. G., Huffaker, D., and Denman, K. Useful techniques of validation for spatially explicit land-change models. *Ecological Modelling* 179, 445, 2004.
21. Train, K. *Discrete Choice Methods with Simulation.* Cambridge University Press, New York, 2003.
22. Webster, D. R. On the edge: Shaping the future of peri-urban East Asia. Stanford University Asia/Pacific Research Center Urban Dynamics Discussion Paper, 2002.
23. Seto, K. C., and Kaufmann, R.K. Modeling the drivers of urban land use change in the Pearl River Delta, China: Integrating remote sensing with socioeconomic data. *Land Economics* 79, 106, 2003.
24. Seto, K. C. et al. Monitoring land-use change in the Pearl River Delta using Landsat TM. *International Journal of Remote Sensing* 23, 1985, 2002.
25. Seto, K. C., and Fragkias, M. Quantifying spatiotemporal patterns of urban land-use change in four cities of China with time series landscape metrics. *Landscape Ecology*, 20, 871, 2005.
26. Seto, K. C. Urban growth in South China: Winners and losers of China's policy reforms. *Petermanns Geographische Mitteilungen* 148, 50-57, 2004.
27. Brennan, G., and Buchanan, J. M. *The Reason of Rules: Constitutional Political Economy.* Cambridge University Press, Cambridge, U.K., 1985.
28. Cohen, B. Urban growth in developing countries: A review of current trends and a caution regarding existing forecasts. *World Development* 32, 23, 2004.

Part III

Synthesis and Prospect

9 Synthesis, Comparative Analysis, and Prospect

Michael J. Hill and Richard J. Aspinall

CONTENTS

9.1 INTRODUCTION

This book has examined the issue of integrated analysis of spatial structure and spatiotemporal processes related to land use change in terrestrial coupled human environment systems. The consequences of land use change have been to transform a large proportion of the land surface of the Earth, and these changes are influencing the global carbon cycle, regional climate, water quality and distribution, and biodiversity through habitat loss.[1] The book presents a number of case studies that include urban, wilderness, wet tropical forest, and arid desert-like environments, human population densities from high to very low, and those that demonstrate both direct and indirect human influences (Table 9.1). The case studies include several that focus on analysis of clearing of tropical forest environments (Chapters 4, 5, 6, and 7; Table 9.1). These are environments of high significance to global carbon stocks, biodiversity, regional climate, and African, Asian, and South American economies.[2,3] However, both intensive agricultural lands (Chapter 3; Table 9.1) and extensive grazing lands (Chapter 2; Table 9.1) are also covered, while specific attention is paid to the late 20th-century phenomenon of the rise of megacities (Chapter 8; Table 9.1).

These case studies and other literature[1,2,3,4,5] suggest that a trade-off approach is probably the only pragmatic option to moderating the impact of humans on the terrestrial system. The pace of modification of the land surface shows no sign of slowing while humans modify their approaches in response to changes in demand and price for product and to avoid detection and control.[6] In this chapter we briefly summarize the main points from each chapter and then examine in turn the main

TABLE 9.1
A Summary of the Key Issues, Contexts, and Methods Examined in Each Chapter of the Book

Chapter	Problem context	Methodological approach(es)	Geographical-system context
1. Aspinall	Technical overview of dynamics, scale, accuracy, uncertainty, pattern and process	Models—conceptual, GIS, RS, CA, MAS, simulation, statistical, empirical, visualization, space-time scaling	International research frameworks—GLP, bio-complexity, etc.
2. Hill	Transformation of spatiotemporal data and relationships to simple indexes	Integration of time-series analysis, spatial analysis, numerical and heuristic methods	Savanna and grassland biomes/livestock as agents
3. Byron/Lesslie	Social surveys of attitudes to natural resource management issues	Assignment of relationships between landholder perception and opinion and biophysical features or management practices	Rural eastern Australia
4. Babigumira, Müller, and Angelsen	Tropical forest clearing	Spatially explicit logistic models	African tropical forests
5. Etter and McAlpine	Tropical forest clearing	Regression tree and regression models of deforestation/regeneration	Amazonian tropical forest
6. Crews-Meyer	Fragmentation of forest	Patch panel metrics—pixel-patch histories	Thailand tropical forest and regrowth
7. Millington and Bradley	Tropical forest clearing	Spatial imprint of cadastral grids; differential behavior of fragmentation metrics as forest cover declines	Amazonian tropical forest
8. Fragkias and Seto	Urban expansion/megacities	Multiple logistic regression, pseudo-Bayesian model averaging to get predicted probabilities of change	Cities of Pearl River Delta—Shenshen, Guangzhou, and Foshan, China

messages described above and look at the potential for delivery of societal benefit from a holistic understanding of, and approach to, land use change.

9.2 A SUMMATION OF THE CHAPTERS

The approach in this book can be summarized in terms of the problem context, methodological approaches, and geographical-system context (Table 9.1). The science context for the theme of the book has been outlined in Chapter 1. Five basic science questions were identified: dynamics of change in space and time; integration of feedbacks between landscape, climate, socioeconomic and ecological systems; resilience, vulnerability, and adaptability of coupled human and natural systems; scale issues; and accuracy and uncertainty issues. Approaches incorporating all of these considerations can be grouped under the general heading of "models," but this includes a diversity of types from conceptual through to empirical (Table 9.1). A broader context for incorporating the different elements of case studies of land use change is to consider a range of integrating frameworks (Figure 9.1). These include the analytical structure for the Global Land Project[7]; the Human Ecosystem Model[8]; the very general framework of explicit focus on linkages between the dynamics of human and natural systems of the U.S. National Science Foundation Biocomplexity in the Environment program (http://www.nsf.gov/geo/ere/ereweb/fund-biocomplex.cfm.), and, with a view to a more design-oriented approach to landscape change, the Landscape Design Research Framework.[9] All of these frameworks seek interdisciplinary definition and focus on key questions within the broad scope of land use change and links to management of change in coupled human environment systems.

In Chapter 2, the role of multiple criteria and trade-off analysis is discussed in the context of methods and approaches for capture and transformation of complex processes. The chapter proposes simple index-based comparative frameworks in an interactive environment that assist decision making, but retain the legacy and critical information content needed for complete appreciation of issues associated with land use change. The analysis seeks to balance evidence-based science and soft-systems approaches by integration of hard data with value judgment, public opinion, and policy and management goals. Temporal analysis is important for entraining legacy and historical factors in decision making; spatial analysis is important for appreciation of social, economic, and biophysical impacts outside the boundary of the directly affected area or geographical location of interest. The approach seeks to measure and aggregate the performance of alternative options. It requires a highly systematic and transparent approach to management of information.

The application of methods that address the science issues identified above, and aggregation of diverse information and data sources into frameworks to assist decision making, must be mediated by the policy context for the analysis and modeling. In the Introduction we suggested that sustainability science, and specifically the tongue model of Potschin and Haines-Young,[10] offers a context and "choice space" for linking science and decision making in policy and management. The case studies provide a mix of analysis across social, economic, and biophysical perspectives necessary to develop an understanding of land use change relevant to sustainability.

In Chapter 3, a social survey approach is used to develop understanding of the attitudes and perceptions of rural communities. The work introduces important methodological issues surrounding relationships between point survey data and spatially explicit biophysical features of a landscape. The analysis looks at relationships between answers to survey questions about land management practices and undesirable landscape features based on a distance analysis. This immediately introduces interesting questions about how humans perceive their spatial environment and how their spatial sensitivities and awareness differ. It also requires some attention to the legacy effects of historical experience and views of the landscape, rooted in historic paradigms, and the impact of aspatial views, norms, and media issues

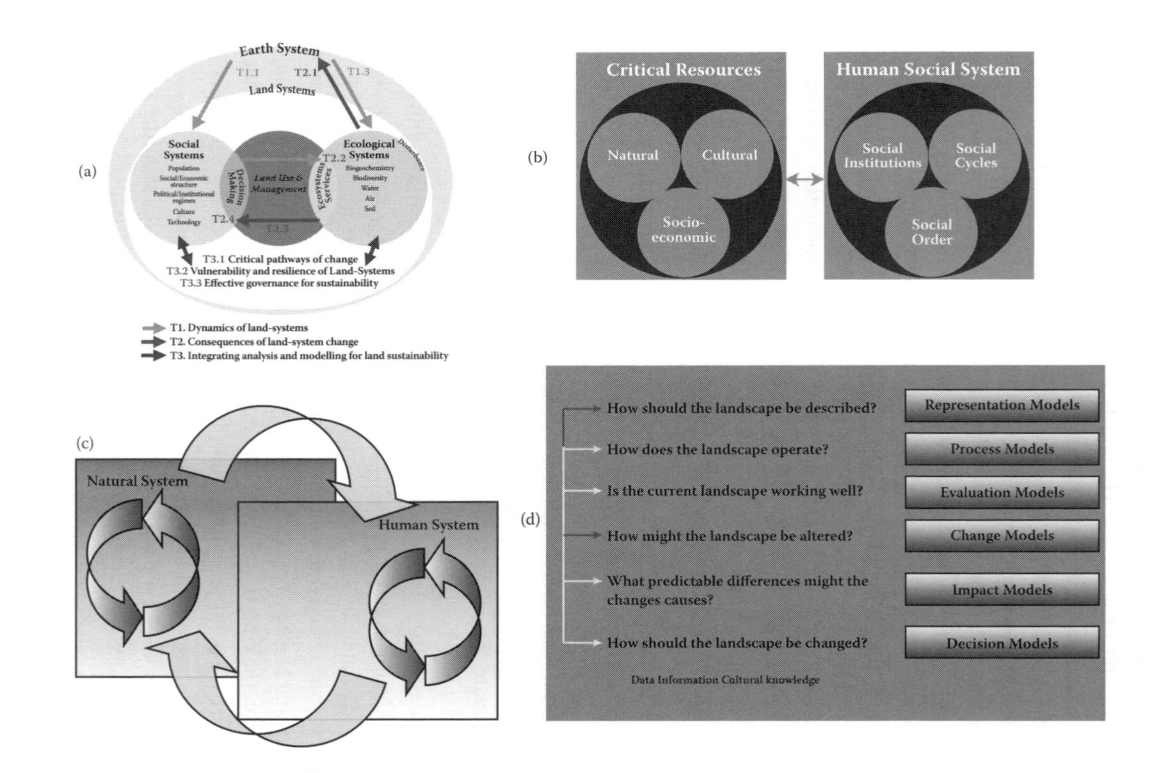

FIGURE 9.1 **(See color insert following p. 132.)** A variety of integrating frameworks that seek interdisciplinary definition and focus on key questions within the broad scope of management of land use change in coupled human environment systems. (a) Analytical framework for the Global Land Project of IGBP and IHDP. (From GLP, 2005, with permission.) (b) The Human Ecosystem Model. (From Machlis et al.,[8] 1997, with permission.) (c) The U.S. National Science Foundation program on Biocomplexity in the Environment (http://www.nsf.gov/geo/ere/ereweb/fund-biocomplexity.cfm); (d) The Landscape Design Research Framework. (From Steinitz et al.,[9] 2003, with permission.)

in society. The chapter is important because it begins to address spatial analysis of opinions, perceptions, and attitudes.

In Chapter 4, an economic analysis centered on the concept of land rent as a driver of land use change is applied to assessment of deforestation in Uganda. An econometric model (binary—logistic) incorporating explanatory variables that describe socioeconomic, spatial, and institutional contexts is used to estimate the probability of deforestation. The analysis tests a set of hypotheses and seeks to define reasons for change. This analysis is also interesting for what it does not include: social survey of attitudes (such as described in the previous chapter) could add important missing motivational information on agent behaviors. The analysis implicitly assumes that motivation will solely be based on maximization of land rent. This is one of the main issues for analysis of coupled human environment systems: is economic return a fully adequate driver for land use change and consequent land cover change in most circumstances? An assessment of the first two case studies might conclude that each would benefit if both of their approaches were combined (i.e., both social survey and econometric modeling were applied). However, evidence from the other case studies shows that further improvement might be obtained if more sophisticated spatial and temporal analysis was applied in conjunction with social survey and economic modeling.

The next three chapters all address deforestation in tropical forests. In Chapter 5, regression tree analysis and logistic regression are applied to analysis of deforestation and regeneration. Social and economic processes are important, as is local and regional context, and deforestation follows a temporally explicit trajectory described by a sigmoidal curve. Patterns are linked to process at different scales: national, regional, and local. Although distances to roads and towns, proximate factors in the terminology of Geist and Lambin,[11,12] are the best predictors at the national level, deforestation and regeneration occur at local hot spots at a regional level. However, at a local level more explicit relationships are obtained with accessibility and soil type, since deforestation rates are strongly related to a spatial metric—forest edge density. This provides an example of spatially explicit analysis deriving a metric with direct meaning in relation to deforestation potential associated with accessibility. Hence, spatial analysis and calculation of pattern metrics can be used to generate indicators of likelihood of deforestation at scales in which pattern and process are directly connected.

This theme of pattern metrics as descriptors or indicators of landscape processes is continued further in Chapter 6. Here, changes in pattern metrics are analyzed for landscape patches through a number of time steps using remotely sensed imagery. In particular the interspersion-juxtaposition index and mean patch size index provide measures of fragmentation. These indices give temporal profiles for different land cover classes such as forest, savanna, and rice agriculture. The analysis in the chapter combines the definition of landscape change in terms of pattern metrics, for example, fluctuations through time in the interspersion of land use/land cover classes, with assignment of meaning to the changes in pattern metrics, for example, reduction in forest interspersion as forest presence declines with the spread of rice agriculture. Explanation of regional differences is based on use of contextual information, for example, proximity to areas of military instability deterring agricultural encroachment. This chapter illustrates the application of sophisticated spatiotemporal analysis combined with the explanatory contextual information, as discussed in Chapters 1 and 2. The quality of this analysis is highly dependent on the sensitivity and accuracy of change detection from remote sensing.

The assessment of forest fragmentation is continued in Chapter 7 where the goal is to increase understanding of relationships between road construction and forest fragmentation in Amazonia. Here, the focus is on the behavior of the agents of change rather than on the structure of the land cover. Hence, if Chapter 6 approaches the issue with a pattern to process orientation, Chapter 7 is examining the same issue with a process to pattern orientation.[13] A six-phase conceptual model of the development of forest fragmentation is described. The approach uses a combination of initial context (colonization) and resulting spatial arrangement of disturbance (the road system) as the foundation to develop the conceptual model, which at each phase has a socio/econo-political context that drives establishment of more elaborate spatially explicit landscape structure, leading to increased fragmentation. Pattern metrics such as mean patch size and total edge length are good descriptors of fragmentation, aligning well with the edge metric correlation with deforestation in Chapter 5, and the interspersion and patch size metric relating to agricultural expansion in Chapter 6. However, the goal to thicken understanding of the human dimension of the change leads to the development of a model with an emphasis on context in the predictive process. This landscape is one of a particular pattern (i.e., herringbone clearance pattern), as a result of the colonization context, which establishes the spatial skeleton, that is the foundation for the final fragmentation pattern.

The three chapters that deal with spatial analysis of tropical deforestation and fragmentation provide a powerful case for combination of spatial analysis and metrication of spatial patterns in disturbed landscapes; analysis of changes in spatial metrics through time as indicators of particular processes and particular trajectories in landscape structure to which economic and biophysical functionality can be ascribed; and application of detailed analysis of socio/econo-political contexts in order to explain evolution of spatial patterns and regional differences in patterns described by spatial metrics.

The final case study in Chapter 8 addresses the issue of rapid urban transformation. A decision theory framework is presented that uses policies, economic data, an economic model, and an optimization procedure that minimizes an objective function to produce probability maps of predicted urban expansion. There is no best or true outcome; the output is a probability surface. The expected rates of growth for the global megacities—there were already 19 cities with populations in excess of 10 million in 2000—introduces an urgency to development of predictive models for planning due to expanded need for energy, sanitation, transport, education, emergency management, health and safety, and clean air and water. Capture of social, psychological, and economic drivers within a complex spatial context is even more important. Many cities are already "landscapes of fate" that have most of the undesirable properties described later in this chapter, and depend upon wealth generation and gentrification of poorer or uglier areas for significant transformation back to a more "desirable" landscape. Therefore, the modeling described in Chapter 8 is of paramount importance in providing spatially explicit probabilities indicating areas for development that may enable some balance to be attained between desirable and fateful urban landscapes for human habitation. The urban case requires more attention to spatially explicit attribution of sociospatial properties and measures of quality of built spatial habitats for human activities in order to balance the powerful influence of city land values and city-based commerce and financial enterprise.[14]

9.3 OVERALL MESSAGES

The framework of the GLP provides a useful template within which to explore the key messages arising from both the overview and case study chapters presented in this book. We have used the elements from Figure 2 of the GLP Science Plan and Implementation Strategy[7] ("The continuum of states resulting from the interactions between societal and natural dynamics") to illustrate the key enabling technologies, methods, and approaches needed to provide real societal benefit from analysis and study of these interactions (Figure 9.2). Simply put, four of the major messages from the case studies are:

1. In order to detect and accurately measure change, remote sensing data and associated methodologies must deliver the highest possible information content and accuracy in change discrimination.
2. Remotely sensed data should be complemented with detailed household and other socioeconomic data from field and census surveys to address decision-making processes in detail and gain a better understanding and capacity to model human and other social and economic processes influencing land use change.
3. Using the basic remote sensing and survey data resources (numbers 1 and 2) together with spatially explicit descriptions of social, financial, jurisdictional, political, and psychological units and influences, there needs to be concerted and integrated application of a variety of numerical, heuristic, spatial, and temporal methods to derive the highest levels of understanding and quantification of dependency between patterns and processes.
4. The analysis from number 3 should be placed in a pragmatic context through closer relations between science with management and policy related to land use and land use change.

9.3.1 Realizing the Full Potential from Remote Sensing

The implementation plan for the Global Earth Observation System of Systems (GEOSS)[15] has defined nine key targets for delivery of societal benefit over the next 10 years, including improving the management and protection of terrestrial and coastal ecosystems; supporting sustainable agriculture and combating desertification; and understanding, monitoring, and conserving biodiversity. A large number of observational requirements have been defined for ecosystems, biodiversity assessment, and agricultural monitoring. These include particular properties associated with land use change such as burned areas, land degradation, species distribution, alien species, extent and location of ecosystems and habitat types, fragmentation of ecosystems and community composition, cultivation and clearing, and grazing

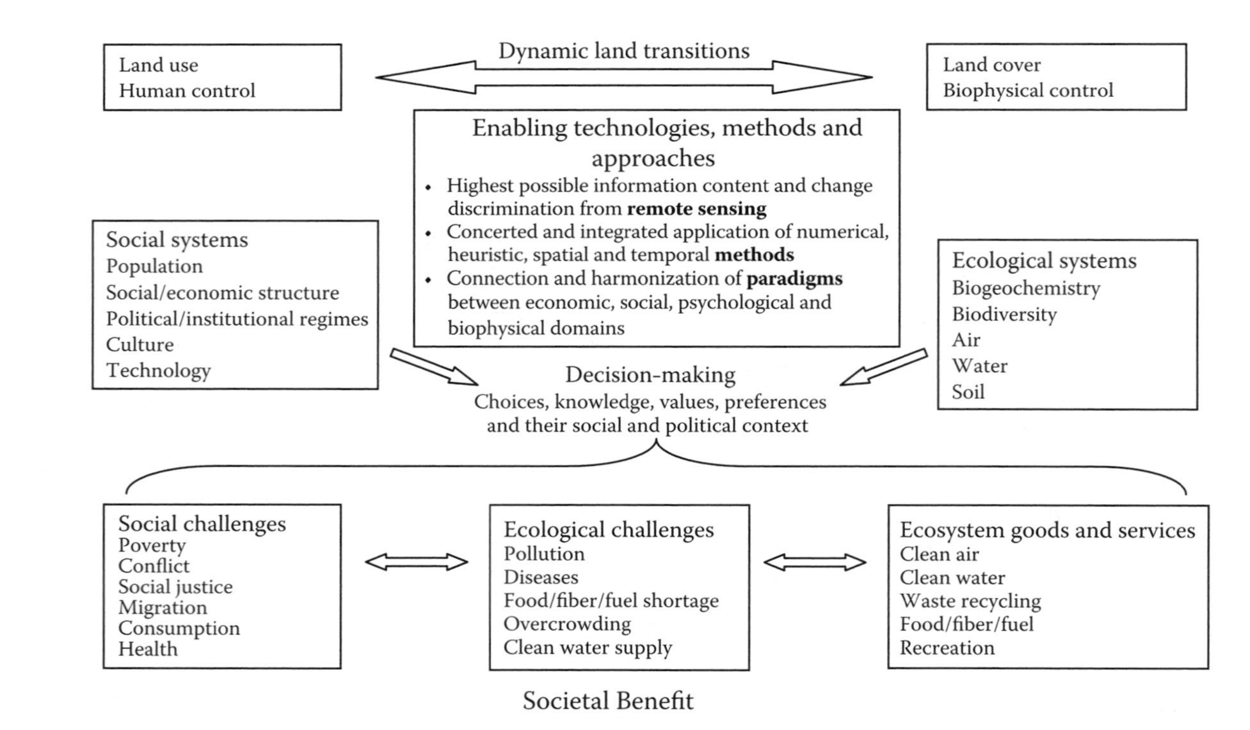

FIGURE 9.2 A diagram of the major elements of the coupled human environment system (after GLP, 2005; Figure 2) augmented with key enabling technologies, methods and data, and paradigms for analysis of coupled human environment systems and delivery of societal benefits.

impacts. An improvement in the quality and coverage of observations of the land surface from remote sensing is needed to realize these requirements.

All of the case studies here that address tropical rain forest clearance (in South America, Asia, and Africa) depend to a large degree on remote sensing as a primary source of data for basic change detection. Crews in Chapter 6 discusses the problems with multiplicative errors when using images from multiple dates, even with high accuracies for individual classifications. The increasing availability of very high-resolution imagery from space (down to 60 cm pixel resolution) means that very detailed definition of land cover boundaries and individual vegetation units is possible at specific locations. However, high image cost and small image footprints mean that this approach is still impractical for widespread change detection. Recent research has shown that detection of changes in forest systems, previously limited by the insensitivity of multispectral instruments such as Landsat to small changes in spectral signatures in heavily foliated forest systems, can be greatly improved using spectral unmixing with time series of multispectral data[16] and with high spectral resolution space-borne sensors such as Hyperion.[6,17] A global hyperspectral imaging system with sufficient signal to noise, moderate pixel resolution (40 to 60 m), and high cycle for global coverage (15 to 30 days) could dramatically increase accuracy and sensitivity of land cover change detection due to land use practices. It is arguable that the natural conclusion to the development of remote sensing technology is the capability to undertake spectroscopy of biospheric surface targets to deliver quantitative values for key surface structural and biogeochemical properties.[18]

9.3.2 Application of Integrated Methods

The application of integrated methods depends on the comprehensive addressing of the basic science questions outlined by Aspinall in Chapter 1. Many accuracy and scale issues can be addressed by maximizing the information content from remote sensing. The information content from remote sensing is maximized by providing a synergistic mix of imagery with full spectral fidelity and calibration, complete global coverage and high temporal frequency at medium resolution, and full spectral fidelity and calibration with very high spatial resolution such that radiometric and spatial scaling is optimized. This remote sensing system would satisfy many of the initial needs of GEOSS,[15] and the products of this system would provide a benchmark level of reliable, spectrally comprehensive, temporal and spatial coverage with explicit quantitative uncertainty estimates.

This improved information content in remote sensing addresses the need for better capture of system dynamics in time and space in the future. However, historical analysis can be greatly enhanced just by comprehensive analysis of global Landsat MSS and Landsat TM/ETM archives. The Australian government supported an innovative program to analyze more than 30 years of Landsat data for Australia in order to monitor and measure land cover change to support a national carbon accounting system.[19,20,21] A large archive exists for the United States, for example, but a full-time series analysis of this has yet to be undertaken (research has commenced to examine forest disturbance using these data but only at a regional scale; Sam Goward, personal communication). The preceding is not intended to overemphasize the importance of remote sensing in integrated analysis, merely to highlight the critical role it can and should play.

The important biophysical processes are described in detail by many models of varying complexity.[22,23] Often these models, developed for site-based application, are difficult to supply with spatially explicit parameters and inputs, resulting in insufficient information to constrain model parameters and provide effective model predictions.[24] The cross-fertilization between disciplines, initiated by Earth system concerns and focus, has provided many numerical, quantitative, and heuristic methods for finding optimal model fits to observations using diverse data with different spatial and temporal properties. These mathematical techniques are collectively referred to as multiple-constraints model-data assimilation.[24] The methods are diverse and have migrated from diverse disciplinary domains such as numerical weather prediction[25] and economic optimization and mathematics.[26,27] These approaches, specifically identified in the Global Land Project strategic implementation plan (Figure 9.1), and in the GEOSS 10-year implementation plan[15] may provide ways to combine information from all available data sets to capture spatially explicit surfaces processes that directly influence land systems and represent many of the consequences of land use change.

The capture of these biophysical processes is one part of the puzzle needed to understand the feedbacks in coupled human environment systems.[28] The other part involves the integration and linking of the spatially and temporally explicit dynamics from observation and biophysical modeling with social and economic measures and metrics, and with cultural views and human perceptions and opinions. The research efforts to "socialize the pixel"[29] have been paralleled by digitizing of parcel information in parts of the United States, for example (thereby providing detailed ownership histories and fine scale land statistics[30]), and increased ability to consider the spatial organization of populations, leading to development of spatially explicit data sets containing survey information on attitudes and behaviors with concomitant concerns for confidentiality.[31] In addition observation of urban ecosystems is explicitly considered with, for example, the Baltimore and Phoenix sites in the U.S. Long Term Ecological Research network.[32]

Integration of all this information and understanding of dynamics may be provided by land use change models[33] that range from cellular automata types,[34] to statistical or simulation models,[33] agent-based models,[35] and integrated ecological and economic modeling.[36] These models must accommodate dynamics across scales, represent the driving forces, capture spatial interactions and neighborhood effects, capture the temporal dynamics, and then be capable of integration across disciplinary domains.[33] Some caution is required here, since we have been down this path with very complex process-based biological and ecosystem models. We may in fact need a "qwerty" solution, a pragmatic combination of methods and approaches that is good enough and that interfaces with the human decision framework, but that is not necessarily the optimum in science or computation—something judged by its effectiveness rather than elegance. However, there is some reason for optimism, expressed earlier, that computational power, process understanding, and diverse methods will deliver highly sophisticated and effective analysis of complex coupled human environment systems. A range of tools that enable humans to interface meaningfully with the information needed for decisions are also required. One discussed briefly in Chapter 2 is multicriteria analysis with a spatial interface designed for interactive work-shopping;[37] however, agent-based models and other interactive paradigms may work just as well.

Through incremental research over many years, cross fertilization of methods between disciplines, digital capture of more spatially explicit biophysical and social data, and continued computational advance, the prospects for sophisticated analysis of complex coupled human environment systems leading to much better informed decision making have become realizable. One example of this process from Australia sees policymakers and scientists driving a highly sophisticated land use mapping process, accompanied by a sophisticated approach to definition of land management practices[38] and development of a robust ecosystem science-based classification system for vegetation disturbance.[39,40] However, the analysis of the coupled human environment system must always be placed in context. The trade-off battle between anthropocentric and envirocentric views of the future of the earth system will be a major battle ground in the 21st century. So context matters.

9.3.3 Placing Analysis in Context

The requirement for delivery of societal benefit that has been emphasized in many forums[7,15] has placed an imperative on the integration and harmonization of analysis, modeling, and decision making and linkage of science to management and policy making by society. Emphasis is placed on the context for problems and whether they are important (or perceived as important). This often depends on communication of issues in a language or framework that impacts directly on society members.

Although not without controversy, the simple societal model of Luhmann[41] provides one way to conceptualize the problem (Figure 9.3). Luhmann views society as a centerless set of "function systems" that constrain both what can be communicated and how it is communicated. He labels economy, law, science, politics, religion, and education as the most important function systems in contemporary society.[42] These function systems have different time intervals for external communication, ranging from daily discourse in science and religion, monthly court processes in law, quarterly teaching and reporting cycles in economics and education, to annual election cycles in politics (average over all levels of government). Information from one function system only becomes active in another function system when it is translated into the code of that function system. Hence, environmental information does not have an impact on economic or political process until it is translated into the code of the economic or political function system. Global climate change is a science/environment issue that has crossed between function systems and is widely considered in the political function system. However, there is still a process of continued transfer of improved scientific information and a continued demand for information in a form that can be used to make political decisions.

If we therefore return to the choice between alternate future landscapes,[10] the "landscapes of desire" and the "landscapes of fate," even if information from environmental science has been translated into the code of the political and economic systems to say that, for example, "if we continue with a certain form of management, certain consequences will ensue", there remains a need to provide the

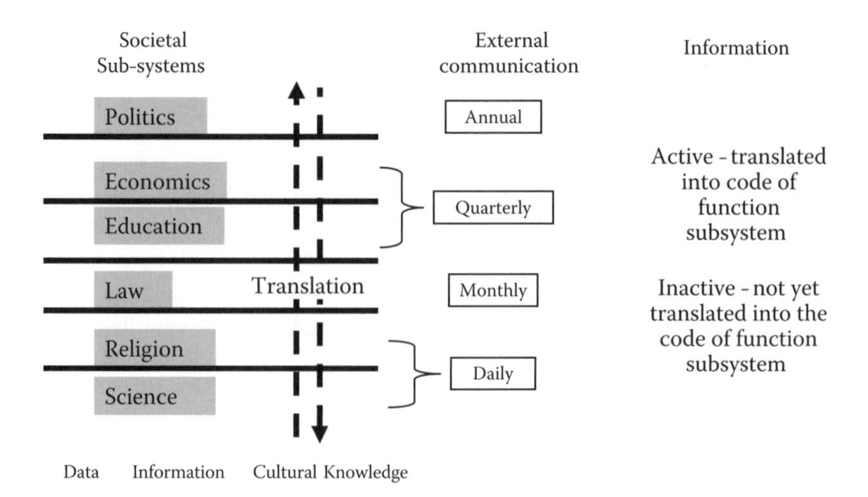

FIGURE 9.3 Connecting the paradigms. Modeling coupled natural human systems. Luhmann[41] views society as a centerless set of "function systems" that constrain both what can be communicated and how it is communicated. He labels economy, law, science, politics, religion, and education as the most important function systems in contemporary society.[42]

methods for excellent analysis of different scenarios and outcomes (Figure 9.4). For example, the number of cities with one million people has grown from 16 in 1900s to more than 400 in 2000, and the number of cities with 10 million people has grown from one in 1950 to 19 in 2000 (see Chapter 8). There is a need to have very sophisticated methods that can explore the wide range of interactions and consequences ensuing from such rapid urban growth. These methods must address the key science questions from Chapter 1 in providing the best possible analysis and scenarios for decision making, since this will inevitably involve "trade-offs between immediate human needs and maintaining the capacity of the biosphere to deliver goods and services in the long term."[1] Since these decision spaces will be highly contested, the quality and transparency of data, methods, and assumptions will be of paramount importance. In addition, regional land use change with negative impacts may have global consequences through climate-surface interactions.[43]

9.4 CONCLUSIONS

This book documents the development of analysis of land use change through presentation of a range of case studies, placed in context through review of land use science, integrative methods, and frameworks for addressing complexity and interrelationships. An abiding theme that lingers subliminally throughout this book is the time-limited decision space of Potschin and Haines-Young[10] and the trajectories of Steinitz and colleagues.[9] By these means human decisions lead to "landscapes of fate" (perhaps a completely urbanized world), and human aspirations crave "landscapes of desire" (perhaps a fairytale land). The challenge is to use the sophisticated science married to social and soft systems paradigms to weave a path to a landscape that preserves the full dimensions of human desire and aspiration while retaining a fully functional earth system.

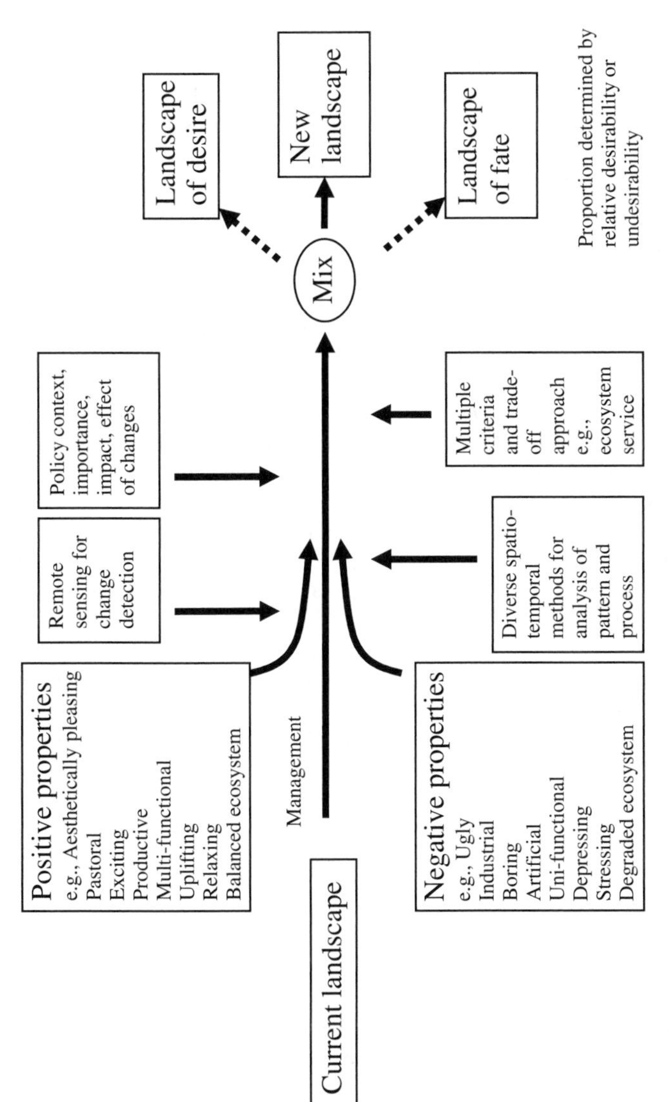

FIGURE 9.4 The context for spatial and temporal analysis in determining the mix of landscapes of fate and desire.

REFERENCES

1. Foley, J. A. et al. Global consequences of land use. *Science* 309, 570–574, 2005.
2. Lewis, S. L. Tropical forests and the changing earth system. *Philosphical Transactions of the Royal Society B* 361, 195–210, 2006.
3. Soares-Filho, B. S. et al. Modelling conservation in the Amazon basin. *Nature* 440, 520–523, 2006.
4. Silva, J. F. et al. Spatial heterogeneity, land use and conservation in the cerrado region of Brazil. *Journal of Biogeography* 33, 536–548, 2006.
5. Defries, R. S., Asner, G. P., and Houghton, R. Trade-offs in land use decisions: Towards a framework for assessing multiple ecosystem responses to land-use change. In: Defries, R. S., Houghton, R., and Asner, G. P., eds., *Ecosystems and Land Use Change*. Geophysical Monograph Series 153, American Geophysical Union, 2004.
6. Asner, G. P. et al. Ecosystem structure along bioclimatic gradients in Hawaii from imaging spectroscopy. *Remote Sensing of Environment* 96, 497–508, 2005.
7. GLP. *Global Land Project. Science Plan and Implementation Strategy.* IGBP Report No. 53/IHDP Report No. 19. IGBP Secretariat, Stockholm, 64 pp, 2005.
8. Machlis, G. E., Force, J. E., and Burch, W. R. The human ecosystem, Part I: The human ecosystem as an organizing concept in ecosystem management. *Society and Natural Resources* 10, 347–367, 1997.
9. Steinitz, C. et al. *Alternative Futures for Changing Landscapes: The Upper San Pedro River Basin in Arizona and Sonora*. Island Press, Washington, D.C., 202 pp, 2003.
10. Potschin, M., and Haines-Young, R. "Rio+10", sustainability science and Landscape Ecology. *Landscape and Urban Planning* 75, 162–174, 2006.
11. Geist, H., and Lambin, E. F. Proximate causes and underlying driving forces of tropical deforestation. *Bioscience* 52, 2, 143–150, 2002.
12. Geist, H. J., and Lambin, E. F. Dynamic causal patterns of desertification. *Bioscience* 54(9), 817–829, 2004.
13. Laney, R. A process-led approach to modelling land use change in agricultural landscapes: A case study from Madagascar. *Agriculture, Ecosystems and Environment* 101, 135–153, 2004.
14. Daniel, T. C. Whither scenic beauty? Visual landscape quality assessment in the 21st century. *Landscape and Urban Planning* 54, 267–281, 2001.
15. GEO. *Global Earth Observation System of Systems. 10-year Implementation Plan and Reference Document.* B. Battrick, ed. European Space Agency Publications Division, Netherlands, 209 pp, 2005.
16. Asner, G. P. et al. Selective logging in the Brazilian Amazon. *Science* 310, 480–482, 2005.
17. Asner, G. P. et al. Drought stress and carbon uptake in an Amazon forest measured with spaceborne imaging spectroscopy. *Proceedings of the National Academy of Sciences* 101, 6039–6044, 2004.
18. Hill, M. J., Asner, G. P., and Held, A. A. Hyperspectral remote sensing of vegetation in coupled human–environment systems—societal benefits and global context. *Journal of Spatial Sciences* 32, 49–66, 2006.
19. Richards, G. P. The FullCAM carbon accounting model: Development, calibration and implementation for the National Carbon Accounting System, National Carbon Accounting System. Technical Report No. 28, Australian Greenhouse Office, Canberra, 2001.
20. Furby, S. Land cover change: Specification for remote sensing analysis. National Carbon Accounting System, Technical Report No. 9. Australian Greenhouse Office, Canberra, 2002.

21. Furby, S., and Woodgate, P. W., eds. Remote sensing analysis of land cover change-pilot testing of techniques. National Carbon Accounting System, Technical Report No. 16, Australian Greenhouse Office, Canberra, 2002.
22. Plummer, S. E. Perspectives on combining ecological process models and remotely sensed data. *Ecological Modelling* 129, 169–186, 2000.
23. Nightingale, J. M., Phinn, S. R., and Held, A. A. Ecosystem process models at multiple scales for mapping tropical forest productivity. *Progress in Physical Geography* 28, 1–41, 2004.
24. Barrett, D. J. et al. Prospects for improving savanna carbon models using multiple constraints model-data assimilation methods. *Australian Journal of Botany* 55, 689–714, 2005.
25. Todling, R. Estimation theory and atmospheric data assimilation In: Kasibhatla, P. et al., eds., *Inverse Methods in Global Biogeochemical Cycles*. American Geophysical Unions, Washington, D.C., 49–65, 2000.
26. Whitley, D. An overview of evolutionary algorithms: practical issues and common pitfalls. *Information and Software Technology* 43, 817–831, 2001.
27. Yang, R. L. Convergence of the simulated annealing algorithm for continuous global optimisation. *Journal of Optimization Theory and Applications* 104, 691–716, 2000.
28. Verberg, P. H. Simulating feedbacks in land use and land cover change models. *Landscape Ecology* 21, 1171–1183, 2006.
29. Rindfuss, R. R. et al. Developing the science of land change: challenges and methodological issues. *Proceedings of the National Academy of Sciences* 101, 13976–13981, 2004.
30. Aspinall, R. J. Modelling land use change with generalized linear models—a multimodel analysis of change between 1860 and 2000 in Gallatin Valley, Montana. *Journal of Environmental Management* 72, 91–103, 2004.
31. VanWey, L. K. et al. Spatial demography special feature: confidentiality and spatially explicit data: Concerns and challenges. *Proceedings of the National Academy of Sciences* 102, 15337–15342, 2005.
32. Grimm, N. B. et al. Integrated approaches to long-term studies of urban ecological systems. *Bioscience* 50, 571–584, 2000.
33. Verburg, P. H. et al. Land use change modeling: current practice and research priorities. *GeoJournal* 61, 309–324, 2004.
34. White, R., and Engelen, G. High resolution integrated modelling of the spatial dynamics of urban and regional systems. *Computers, Environment and Urban Systems* 24, 383–400, 2000.
35. Barreteau, O., and Bousquet, F. SHADOC: A multi-agent model to tackle viability of irrigated systems. *Annals of Operations Research* 94, 139–162, 2000.
36. Vionov, A. et al. Patuxent landscape model: integrated ecological economic modelling of a watershed. *Environmental Modelling and Software* 14, 473–491, 1999.
37. Hill, P., Cresswell, H., and Hubbard, L. Spatial prioritisation of NRM investment in the West Hume area (Murray CMA region). Technical Report No. 2006, CSIRO Water for a Healthy Country National Research Flagship, Canberra, 2006.
38. Lesslie, R., Barson, M., and Smith, J. Land use information for integrated natural resources management—a coordinated national mapping program for Australia. *Journal of Land Use Science* 1, 45–62, 2006.
39. Thackway, R., and Lesslie, R. Vegetation assets, states and transitions (VAST): Accounting for vegetation condition in the Australian landscape. BRS Technical report, Bureau of Rural Sciences, Canberra, 2005.
40. Thackway, R., and Lesslie, R. Reporting vegetation condition using the vegetation, assets, states and transitions (VAST) framework. *Ecological Management and Restoration* 7, s53–s62, 2006.

41. Luhmann, N. *Ecological Communication.* Translation Bednarz, J. Jr. University of Chicago Press, Chicago, 187 pp, 1989.
42. Grant, W. E., Peterson, T. R., and Peterson, M. J. Quantitative modelling of coupled natural/human systems: simulation of societal constraints on environmental action drawing on Luhmann's social theory. *Ecological Modelling* 158, 143–165, 2002.
43. Asner, G. P., and Heidebrecht, K. B. Desertification alters ecosystem-climate interactions, *Global Change Biology* 10, 1–13, 2004.

Index

D

E

F

G

H

I

L